CREDIT CORRELATION
Life After Copulas

CREDIT CORRELATION
Life After Copulas

editors

Alexander Lipton
Andrew Rennie

Merrill Lynch International, UK

World Scientific

NEW JERSEY · LONDON · SINGAPORE · BEIJING · SHANGHAI · HONG KONG · TAIPEI · CHENNAI

Published by

World Scientific Publishing Co. Pte. Ltd.

5 Toh Tuck Link, Singapore 596224

USA office: 27 Warren Street, Suite 401-402, Hackensack, NJ 07601

UK office: 57 Shelton Street, Covent Garden, London WC2H 9HE

British Library Cataloguing-in-Publication Data
A catalogue record for this book is available from the British Library.

ISBN-13 978-981-270-949-3
ISBN-10 981-270-949-5

Typeset by Stallion Press
Email: enquiries@stallionpress.com

Printed in Singapore.

CONTENTS

INTRODUCTION

Recent growth of credit derivatives has been explosive. The global credit derivatives market grew in notional value from \$1 to \$20 trillion dollars from 2000 to 2006. However, understanding the true nature of these instruments still poses both theoretical and practical challenges. For a long time now, the framework of Gaussian copulas parameterised by correlation, and more recently base correlation, has provided an adequate, if unintuitive, description of the market. However, increased liquidity in credit indices and index tranches, as well as proliferation of exotic instruments, such as forward starting tranches, options on tranches, leveraged super senior tranches, and the like, have made it imperative to come up with models which describe market reality better.

In view of this fact, Merrill Lynch decided to organize a conference entitled "Credit Correlation: Life after Copulas" in order to discuss the current developments in modelling for credit derivatives and the practical implications. This conference took place in September, 2006 at the Merrill Lynch Financial Centre in the City of London and brought together both practitioners and prominent academics. Practitioners presented 6 talks, while 4 academics participated in a panel discussion. This book volume, reprinted from the *International Journal of Theoretical and Applied Finance* (Vol. 10, No. 4), brings together these talks and panel presentations. The book contains 8 papers (the presentation by Prof. L. C. G. Rogers of Cambridge will be published elsewhere).

All participants agreed that base correlation has outlived its usefulness; opinions of how to replace it, however, were divided. Both the top-down and bottom-up approaches to describing the dynamics of credit baskets were presented and pro and contra arguments were put forward. Proponents of the top-down approach presented several complementary methods for studying the evolution of the loss distribution for a credit basket. Advocates of the bottom-up approach used both reduced-form and structural models. It is fair to say that neither camp won over the other; we leave it to the reader to decide which direction is the most promising at the moment. However, there was a real sense of forward movement and genuine belief in the importance of modelling. We can hope that in the near future, models which transcend base correlation will be proposed and accepted by the market.

Alexander Lipton, Andrew Rennie
Merrill Lynch
1 May 2007

LÉVY SIMPLE STRUCTURAL MODELS

MARTIN BAXTER

Nomura International plc, 1 St Martin's-le-Grand
London EC1A 4NP, United Kingdom
work@martinbaxter.co.uk

This paper considers credit portfolio models based on Levy processes in general, and the gamma model in particular. It describes both single-name and multi-name situations using the gamma model, along with calibration fits and a comparison of various simple Levy models. There is also extensive historical data, including the May 2005 Auto crisis, which can be described in terms of the model. Parameter-based risk management using the gamma model is also discussed along with implementation details.

Keywords: Structural credit model; CDO pricing; Levy process; gamma process.

1. Introduction

The Gaussian copula model does not fit the credit market, and the base correlation method of forcing it to fit has several disadvantages. Its main drawbacks are a theoretical possibility of arbitrage; actual arbitrage in practice; and difficulties in extending the method to price bespoke baskets and more exotic products.

In this paper we will propose a family of Levy structural models which fit the market better than the Gaussian copula. These models are intuitive, easy to implement, and provide insights into both risk management and relative value opportunities. They also extend to bespoke baskets and more exotic products.

The economic idea behind the model is that the tails of the Gaussian distribution are too thin to model the credit market accurately. Although the Gaussian distribution is widely used in other asset classes, it is rarely suitable for extreme out-of-the-money options. But almost all credit default events are extreme events which are controlled by the tail of the distribution. For this reason we reject Brownian motion and look for alternative stochastic processes which have heavier tails.

As movements in the credit market are sometimes sudden and jump-like, we choose to work with Levy processes. These can have both jumps and heavier tails, as we desire. We will also show a way of creating multi-variate Levy processes, which will allow us to break up an entity's jumps into global jumps and idiosyncratic jumps. This mirrors the reality of market movements — May 2005 was a global jump and Parmalat was an idiosyncratic jump.

There has been growing interest recently in studying jump models of individual names. Joshi and Stacey [5] developed a method with a global time-scale factor (intensity gamma) and they achieve CDO calibration. An economic drawback is that it does not allow idiosyncratic spread jumps and its practical limitation is that it requires Monte Carlo. Schoutens [10] drives the intensity as a gamma OU process with spread up-jumps, and has success with CDS calibration, but not CDO calibration. Luciano and Schoutens [6] have a multivariate gamma process also using a global time-change clock. This also has no idiosyncratic spread jumps, and does not match CDO prices.

2. Levy Processes

Let us give a brief summary of the basic properties of a Levy process. Winkel [11] gives a brief introduction, and Applebaum [1] is a very useful reference for Levy processes. Levy processes are, in some sense, a generalization of Brownian motion to contain drift, Brownian motion, and jump terms. Formally a stochastic process $X(t)$ is a Levy process if

- $X(0) = 0$,
- X has stationary increments, that is $X(s+t) - X(s) \overset{d}{=} X(t)$,
- X has independent increments, that is $X(s + t) - X(s)$ is independent of $(X(u) : u \leq s)$, and
- (technical continuity condition), $\lim_{t \to s} P(|X_t - X_s| > \varepsilon) = 0$, for all positive ε.

Given sufficient boundedness on X, its moment generating function can be written as

$$E(\exp(\theta X_t)) = \exp(t\psi(\theta)),$$

where the *Levy symbol* of X is $\psi(\theta)$ which can be decomposed as

$$\psi(\theta) = \mu\theta + \frac{1}{2}\sigma^2\theta^2 + \int_{-\infty}^{\infty} (\exp(\theta x) - 1)\nu(x)\, dx.$$

Its three terms correspond to:

- Constant drift, where μ is the drift coefficient
- Brownian motion, where σ is the volatility
- Pure-jump term, where the *Levy measure* ν gives the intensity of the Poisson arrival process of jumps of size x. That is, jumps whose size lies in $[x, x + dx]$ occur as a Poisson arrival process with intensity $\nu(x)dx$.

A useful basic Levy process is the *gamma process*, which is a pure-jump increasing process with Levy measure

$$\nu(x) = \gamma x^{-1} \exp(-\lambda x), \quad x > 0, \quad \text{and} \quad \text{Levy symbol } \psi(\theta) = -\gamma \log(1 - \theta/\lambda).$$

The gamma process has marginal distributions which follow the (continuous) gamma distribution. Its parameters are gamma (γ) which controls the jump intensity, and lambda (λ) which controls the inverse jump size. We notate it as $X(t) = \Gamma(t; \gamma, \lambda)$.

Another process of interest is the *variance gamma process* which is the difference of two gamma processes

$$X(t) = VG(t; \gamma, \lambda_d, \lambda_u) = \Gamma(t; \gamma, \lambda_u) - \Gamma(t; \gamma, \lambda_d).$$

This process was first used in finance by Madan *et al.* [7], and for modeling credit by Moosbrucker [8].

3. Credit Models for Single Names

We begin with a simple structural model for a single credit name. We define the value of the firm, or a proxy for it, as the log-gamma process

$$S_t = S_0 \exp(-\Gamma(t; \gamma, \lambda) + \mu t), \quad \text{where } \mu = \gamma \log(1 + \lambda^{-1}).$$

Thus we are assuming that $S(t)$ is a positive martingale with up-drift and down-jumps. This simple model assumes that "no news is good news". We also assume that the entity defaults when $S(t)$ goes below a threshold c.

We note two extreme cases. As gamma tends to infinity, and $\lambda \sim \sqrt{\gamma}/\sigma$, then $S(t)$ tends to the log-normal process

$$S_t = S_0 \exp\left(\sigma W_t - \frac{1}{2}\sigma^2 t\right).$$

At the other extreme, as gamma tends to zero, and $\lambda \sim \exp(-h/\gamma)$, then $S(t)$ becomes the constant default-intensity model

$$S_t = S_0 \exp(ht) I(\tau > t), \quad \text{where } \tau \sim \exp(h), \text{ the exponential default time.}$$

These two cases represent the archetypal extremes of credit defaults. In the log-normal case, spreads widen continuously until default, giving warning of the impending default. (Argentina is a relevant example.) In the second case, spreads never change and defaults happen with no warning at all. (Parmalat was more like this.) The gamma parameter lets us model a situation somewhere in between these extremes.

For calculating the default probability we need to evaluate

$$p_T(c) := P\left(\inf_{t \le T} S_t < c\right) = P\left(\sup_{t \le T}(\Gamma(t; \gamma, \lambda) - \mu t) > k := \log(c)\right).$$

There are two ways of doing this. The simplest is just to use the European approximation:

$$p_T(c) \cong P(X_T > k), \quad \text{where } X_t = \Gamma(t; \gamma, \lambda) - \mu t.$$

A more sophisticated approach is to estimate the actual barrier probability using numerical methods. This is more computationally intensive than the European approximation, but can be used in tests to monitor the accuracy level of the European approximation.

The European approximation will not give identical results to the barrier-style method for the same parameters. Instead we would like to be able to calibrate the European-style method so that the European method with European-style parameters closely matches the barrier-style method using barrier-style parameters. If that could be done, then we could have confidence that the European-calibrated European-style prices were a good match for the actual barrier prices of trades.

3.1. *Example: Term structure of a single credit*

Figure 1 shows the 10 y term structure of credit default swap spreads under three different modeling assumptions. Parameters have been chosen to match approximately the shape of the CDX 125 S7 index spreads as at 24 November 2006. The curves show:

1. Gamma model, with barrier pricing method, and (barrier-calibrated parameters) of γ at 49%, and λ at 3.18.
2. Gamma model, with European pricing method, and (European-calibrated parameters) of γ at 49%, and λ at 1.50.
3. Brownian motion barrier model, $X(t) = W(t) + \mu t$, defaulting when $X(t)$ hits a barrier, with μ set to 0.153.

The basket's index spreads are also shown. We have deliberately constrained the important gamma parameter in the European-calibration to equal the barrier-calibration's gamma, but allowed lambda to calibrate separately.

In Fig. 1, we see that the European and barrier versions of the gamma model are very similar, and are a rough fit to the actual market curve. The Brownian curve

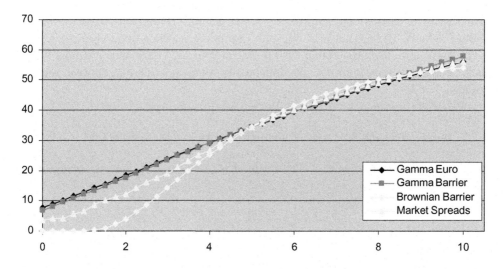

Fig. 1. Credit spread term structure.

also approximates the market curve, but it underestimates the market spreads at the short end. We do not need the fit to the market curve to be exact, as we will allow the threshold to have a term structure, but it is important that the fit is reasonable to justify the basic form of the model.

For this paper, we will use the European approximation throughout, but it is possible to rework the results with the slower barrier formula. We will also check the barrier formula on CDO prices below.

3.2. *Extensions*

We can extend the basic model in a number of ways.

1. We can add "good news" jumps by using the variance gamma process instead of the gamma process.
2. We can add a global catastrophe term with a low-intensity high-impact global factor (such as a Poisson process or other gamma process).
3. We can add a Brownian motion term to get continuous random movement as well.

We can also add various combinations of these, such as Brownian–variance–gamma, catastrophe–variance–gamma, and so on. In fact any Levy process can be used in our CDO modeling, as we shall now describe.

4. Portfolio Credit Models

We need to extend our single-name model to a multi-variate correlated model of a portfolio of names. To do this, we need a way of generating multi-variate Levy processes with a given marginal distribution.

Lemma 4.1. *Multivariate Levy construction. For any Levy process $X(t)$, any integer n, and any non-negative correlation ϕ, we can construct a set of n Levy processes $X_1(t), X_2(t), \ldots, X_n(t)$, such that each $X_i(t)$ has the same distribution as $X(t)$, and the correlation between $X_i(t)$ and $X_j(t)$ is ϕ for all distinct i and j.*

Proof. Start by making $n+1$ independent copies of $X(t)$, called $X_g(t)$, $\tilde{X}_1(t), \ldots, \tilde{X}_n(t)$. Then define

$$X_i(t) = X_g(\phi t) + \tilde{X}_i((1 - \phi)t).$$

Then, by the stationarity and independent-increment properties of Levy processes, it is immediate that $X_i(t)$ is a Levy process with the same distribution as $X(t)$. If $X(t)$ has second moments, then

$$Var(X_i(t)) = \sigma^2 t, \quad \text{for some } \sigma, \text{ and hence}$$
$$Cov(X_i(t), X_j(t)) = \sigma^2 \phi t, \quad \text{so } Corr(X_i(t), X_j(t)) = \phi.$$

An alternative construction is that we construct the global factor $X_g(t)$ as a Levy process with Levy symbol ψ_g, and $\tilde{X}_i(t)$ as IID Levy processes with Levy symbol ψ_i , where

$$\psi_g(\theta) = \phi\psi(\theta), \quad \psi_i(\theta) = (1 - \phi)\psi(\theta), \quad \text{and} \quad X_i(t) = X_g(t) + \tilde{X}_i(t).$$

Even if $X(t)$ does not have second moments, this representation has the interpretation that a fraction ϕ of movements of a single name are due to global effects which affect all other names too. □

Given this lemma, we can formulate our CDO model as follows.

We construct a correlated set of state variables from some independent gamma processes as

$$X_i(t) = -\Gamma_g(t; \phi\gamma, \lambda) - \Gamma_i(t; (1 - \phi)\gamma, \lambda).$$

This decomposes the log-value of the entity into a global and an idiosyncratic gamma process. We assume, as before, that the entity has defaulted by time t, if $X_i(t)$ is below a threshold. This threshold can be calibrated to match precisely the survival probability of the entity to that date.

We note that the lambda parameter is redundant, due to scaling. So the gamma model has only two parameters — gamma and phi.

As an aside, we can reformulate the Gaussian copula as a combination of global and idiosyncratic normal random variables as

$$X_i = \sqrt{\rho}Z_g + \sqrt{1 - \rho}Z_i.$$

This itself can be rewritten as a sum of time-changed Brownian motions as

$$X_i(t) = W_g(\rho t) + W_i((1 - \rho)t),$$

which matches our general Levy multivariate structure.

5. Calibration and Model Comparison

Our general calibration procedure is to take the market's tranche capital structure for a given liquid basket and optimise the model's parameters to achieve the best fit.

The objective function we used is the root mean-squared error,

$$V(\gamma, \phi) = \left(\frac{1}{n} \sum_{tr \neq eq} (Mkt_{tr} - Model_{tr})^2 \right)^{1/2},$$

where the sum is taken over the n non-equity tranches of the capital structure, and $Mkt(tr)$ and $Model(tr)$ are respectively the market and model par spreads for tranche tr. This scheme uses a simple weighting, which ignores equity, but other weighting schemes can be used to redistribute the pattern of fitting errors.

The results of the calibration for the CDX 125 S7 basket, as at 28 November 2006, are shown in Table 1.

The best fit score is also shown (the root-mean-square error in basis points), which excludes the equity error. The general fitting quality is quite good, but not perfect. The larger errors are in

- equity tranches, which are excluded from the fitting objective. Especially in 7y which has a 4% error. This can be reduced by including the equity tranche in the objective function, at the expense of the other tranches.
- senior and super-senior. The model's ratio of the senior spread to the super-senior spread has values 35, 15 and 9, compared with the market's values of 2, 3 and 4. We have not been able to find any model which comes close to these ratios of market prices. Those market price ratios are close to implying an arbitrage opportunity, as buying senior protection and selling (more) super-senior protection will make money for many plausible scenarios on recovery rates and conditional default probabilities.

The equivalent results for iTraxx 125 S6, also at 28 November 2006, are shown in Table 2.

The fitting behaviour is similar to CDX.

The parameter values which achieved these fits are shown in Table 3.

Table 1.

Tranche	5y CDX S7		7y CDX S7		10y CDX S7	
	Market	Model	Market	Model	Market	Model
0%–3%	24.6%	26.6%	40.6%	44.5%	51.1%	52.7%
3%–7%	91.0	90.5	210.0	210.2	426.0	427.2
7%–10%	18.4	19.4	46.8	46.1	110.0	109.1
10%–15%	6.5	7.3	19.0	20.2	51.5	54.0
15%–30%	3.1	1.6	6.0	5.9	14.8	21.0
30%–100%	1.4	0.04	2.3	0.4	3.9	2.4
Fit score (bp)		**1.1**		**1.1**		**3.2**

Table 2.

Tranche	5y iTraxx S6		7y iTraxx S6		10y iTraxx S6	
	Market	Model	Market	Model	Market	Model
0%–3%	14.0%	15.5%	28.9%	30.9%	42.4%	46.0%
3%–6%	63.0	62.6	144.0	144.0	345.5	346.9
6%–9%	17.3	17.7	42.5	42.5	109.0	106.8
9%–12%	7.0	7.9	21.3	20.8	47.8	51.1
12%–22%	2.9	2.7	7.3	8.1	15.0	19.6
22%–100%	1.2	0.1	1.9	0.6	3.4	1.3
Fit score (bp)		**0.7**		**0.7**		**3.0**

Table 3.

	CDX Gamma	CDX Phi	iTraxx Gamma	iTraxx Phi
5y	152.5%	8.9%	135.5%	9.4%
7y	49.2%	10.6%	60.3%	11.5%
10y	15.5%	19.5%	45.8%	13.7%

These parameters have the typical behavior of decreasing gamma with maturity, and increasing phi. Gamma levels are also relatively high in November 2006, which we shall discuss below.

We can also perform the calibration using the barrier formula, rather than the European approximation. In this case the 5y CDX fitted spreads are: 25.9%, 90.5, 19.4, 7.2, 1.5, and 0.04. These are almost identical to the European fitted values, though with slightly different calibrated parameters (gamma 158%, phi 10%, lambda 50%), so the net effect of the barrier formula is similar to a small change in co-ordinates rather than different model prices.

We have also performed a comparison test of a variety of Levy process models. The models used were: the basic gamma model (down jumps only); variance gamma (asymmetric up and down jumps); Brownian gamma (down jumps plus diffusion); Brownian variance gamma (symmetric up and down jumps plus diffusion); catastrophe gamma (down jumps plus global catastrophe factor); catastrophe variance gamma (symmetric up and down jumps plus catastrophe). Additionally the Gaussian copula, without base correlation, was included for reference.

All the models were calibrated to historical market data over the period 12 October 2005 to 5 April 2006, using one date per week. see Table 4.

The average pricing error along the capital structure (these runs included the equity tranche) is shown for each basket/maturity combination, and the total average is in the rightmost column.

The Gaussian copula, as expected, performs badly with an average spread error of 62bp. All the Levy based models are significant improvements on the Gaussian copula. The Brownian–variance–gamma and catastrophe-variance-gamma, which both have symmetric up and down jumps, are relatively poor performers. This

Table 4.

Model	CDX 5y	CDX 7y	CDX 10y	iTraxx 5y	iTraxx 7y	iTraxx 10y	Average (bp)
Catastrophe Gamma	1.4	7.9	15.4	1.1	7.0	8.7	6.9
Variance Gamma	2.9	9.6	15.7	2.9	9.6	7.0	8.0
Gamma	**3.3**	**7.7**	**17.2**	**3.2**	**6.8**	**17.0**	**9.2**
Brownian Gamma	4.7	11.1	18.3	3.9	9.2	13.8	10.2
Brownian Var Gamma	2.8	21.9	44.2	2.3	18.2	40.6	21.7
Cat Var Gamma	1.4	28.6	48.1	1.0	26.4	34.7	23.4
Gaussian Copula	38.9	66.1	76.3	33.6	75.7	83.9	62.4

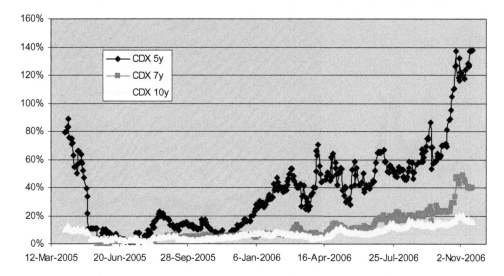

Fig. 2. CDX gamma.

suggests that it is important that up-jumps be different to (and smaller than) the down-jumps.

Interestingly, the best four models have quite similar scores, lying in the range 7–10 bp. Of these, the gamma model has the fewest parameters and greater parameter stability. This parsimony encourages us to focus on the gamma model as our favored model at this stage.

We can calculate the historical gamma and phi parameters for the gamma model. We have data which lets us do this daily from 1 April 2005 up to 20 November 2006. The gamma parameter is shown in Fig. 2, the phi parameter is shown in Fig. 3.

We see that the auto crisis of May 2005 was linked to a crash in the value of gamma from about 80% down to 10%. Since then, gamma has been gradually increasing, with a faster increase in November 2006, which some market commentary called the "reverse correlation crisis" [3].

Phi was relatively stable during the auto crisis, though noisier shortly afterwards, and it trended downwards in the second half of 2005. During 2006 phi has been very stable.

6. Parameter Risks and Hedging

Risks are divided into credit spread risks and correlation skew risks. Because the model is "bottom up", it readily shows credit spread risk (and recovery rate risk) to every name in the basket. This compares with "top down" approaches, such as Brigo *et al.* [2], which focus on the loss distribution and cannot easily give the credit risk of each name.

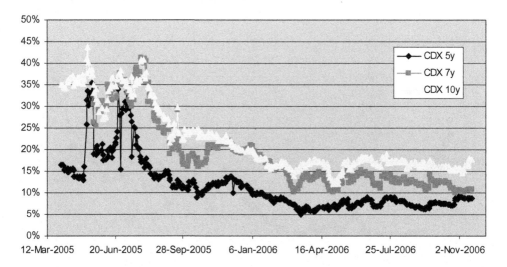

Fig. 3. CDX phi.

Table 5.

	d/dGamma	d/dPhi
0%–3%	−799	−50,532
3%–7%	1,384	15,270
7%–10%	135	8,142
10%–15%	−87	4,423
15%–30%	−110	1,612
30%–100%	−20	137

The correlation risk is given by the risks to the parameters. We can see the parameter risks of the CDX 125 S7 basket along its capital structure. Table 5 shows the risk, tranche by tranche, for a $10 million buy-protection par 5y trade to a 1% increase in gamma and phi levels respectively.

We see that when gamma increases, then the equity and senior spreads decrease and the mezz spreads increase. Increasing gamma moves value from the edges of the capital structure into the middle.

When phi increases, equity spreads decrease and the mezz and senior spreads increase. Increasing phi moves value from the very junior tranches to the more senior.

In base correlation terms, increasing gamma flattens the base correlation curve and increasing phi moves the base correlation curve upwards. A useful interpretation is that phi controls the average level of the base correlation curve, and gamma controls its slope. Indeed, as we saw earlier, an infinite gamma corresponds to the Gaussian copula and a perfectly flat base correlation curve.

For risk management of a CDO portfolio we can hedge gamma and phi (by maturity) to flatten our risk. This is an advance on base correlation, where hedging is based on tranche-by-tranche inventory management. Having a model allows parameter-based hedging, which is similar in spirit to the hedging of stochastic vol parameters in the interest-rate skew market. It also allows us to hedge different parts of the capital structure against each other. We can also price bespoke baskets using IG parameters (adapted as necessary), and hedge the risk with IG tranches.

6.1. *Case study: Auto crisis May 2005*

Prior to the auto crisis, many hedge fund investors had a "positive carry" trade, which involved selling equity protection and buying mezz protection. This was about flat in credit spread risk, and long correlation. It was not appreciated at the time, but this was also a long gamma position, as we can see from the risk table above. The crisis was marked by a general spread widening and a strong gamma sell-off (see gamma history plot above). This widened equity spreads (which are short gamma), but tightened mezzanine spreads (which are long gamma). So the investors (who were long gamma) lost money.

Hindsight makes it easy to make the correct decisions in retrospect, but the gamma model could have provided some risk management information beforehand. It would have warned potential investors both that they were running an exposed position in correlation skew (gamma), and also that gamma was trading at high levels.

7. Implementation and Other Products

The models presented here can be implemented in a similar way to many existing Gaussian copula implementations. In particular, they do not require Monte Carlo simulation, though it is possible to use it.

Let us formulate the Gaussian copula model in a similar way to our Levy process dynamics as

$$X_i(t) = W_g(\rho t) + W_i((1 - \rho)t),$$

where $W(g)$ and $W(i)$ are global and idiosyncratic Brownian motions. Thus X_i is also a Brownian motion. Let us write $F(x;t)$ for the common distribution function of $X(t)$, $W_g(t)$, and $W_i(t)$. A sketch of the implementation could run as follows:

1. For each time t, calculate the threshold $\theta_i(t) = F^{-1}(p_i(t); t)$, where $p_i(t)$ is the default probability of entity i by time t.
2. Integrate over the possible values of the global factor $W(g)$, which has distribution function $F(x; \rho t)$. We can use either simple methods such as Simpson integration, or more sophisticated quadrature techniques. Both methods use an

approximation of the form

$$E(payoff(W_g)) \cong \sum_{k=1}^{m} \alpha_k E(payoff|W_g = y_k),$$

where $y(k)$ are a discrete set of values of $W(g)$, and $\alpha(k)$ are some weights.

3. Given that the global factor $(W(g) = y)$, we calculate the conditional default probabilities of each entity

$$p_i(y, t) = P(X_i(t) \le \theta_i(t)|W_g = y) = F(\theta_i(t) - y; (1 - \rho)t).$$

4. With these conditional default probabilities, we calculate the conditional expectation of the payoff. This is helped by the conditional independence of the entities' values given $W(g)$. The expectation can be performed by approximations such as a normal-approximation to the basket loss, or the ingenious bucket algorithm of Hull and White [4].

To change from the Gaussian copula to our new Levy models, all we have to do is replace the distribution function F used in steps $1, 2$, and 3. Step 4 is unaltered. So the problem reduces to calculating the marginal distribution function for the various models we have used.

7.1. Calculating the distribution function

Gamma model. The gamma model has marginal gamma distributions. Their distribution is already well approximated. See, for example, section 6.2 of *Numerical Recipes* [9]. Quadrature integration against a gamma random variable is also possible, as implemented in routine **gaulag** of *Numerical Recipes* section 4.5. Run-time performance for the Gaussian copula and the gamma model should be broadly similar.

For calculating the inverse of the distribution function, it is effective to perform interval bisection to bracket the root initially, since the distribution function is monotonic, and then apply some Newton–Raphson iterations to polish it.

Variance gamma model. There is a time-change representation of the variance gamma process as

$$VG(t; \gamma, \lambda_d, \lambda_u) = W(A_t) + \mu A_t,$$

where $W(t)$ is a Brownian motion, $A(t)$ is a gamma $\Gamma(t; \gamma, \frac{1}{2}\lambda_d\lambda_u)$ process, and μ is the drift $\frac{1}{2}(\lambda_d - \lambda_u)$. Thus $P(VG(t; \gamma, \lambda_d, \lambda_u) \le x) = P(Z\sqrt{A_t} + \mu A_t \le x)$. The probability can then be expressed as an integral conditional on the value of $\Gamma = A(t)$ as

$$P(Z\sqrt{\Gamma} + \mu\Gamma \le x) = \int_0^\infty f_\Gamma(y)\Phi\left(\frac{x - \mu y}{\sqrt{y}}\right) dy.$$

This integral can be performed efficiently using the gamma quadrature integration mentioned above. Numerical difficulties may occur when the ratio λ_u/λ_d is extreme (larger than 10), so these cases might be excluded.

Other models, such as Brownian gamma, can be handled in similar ways.

7.2. *Performing the optimization*

Gamma model. The optimization for the gamma model is relatively straightforward. There are only two parameters (gamma and phi) and they both have nontrivial and different effects on the tranche spreads. We use an optimiser similar to the Levenberg–Marquardt method, described in section 15.5 of *Numerical Recipes*. About half-a-dozen iterations are enough to get a good calibration.

Other models. We use the same optimizer, but the situation is more complicated. The function mapping parameters to tranche spreads is significantly nonlinear and the presence of semi-redundant parameters increases the difficulty. For such difficult parameters, we try optimizing whilst keeping that parameter fixed, and then we vary the parameter and optimize again. This is slow but effective.

7.3. *Other products*

The model can be extended to price bespoke tranches, bespoke baskets and more exotic products. Bespoke tranches and tranchelets are priced immediately in the same way as standard tranches above. Bespoke basket can also be priced, once we know the values of gamma and phi for the bespoke basket. Those, as ever, have to be estimated from the liquid basket parameters. Investment-grade parameters can be estimated as the average of the CDX and iTraxx parameters, and the high-yield basket CDX HY is also liquid giving an estimate of high-yield parameters. A straightforward bespoke pricing scheme is just to take a convex combination of the IG and HY parameter sets, driven by the spread of the bespoke basket. Bespoke pricing is never certain, so room still remains for trading judgement on a bespoke basket's gamma and phi levels. So uncertainty is reduced down to two simple parameters.

More exotic products, such as CDO-squared and long-short CDO, can also be priced under the model. By conditioning on the global factor, the product can be priced using the conditional independence of the names.

8. Summary and Conclusions

We have presented a family of Levy process models for single-name credits and baskets. Of these, a simple and effective model is the gamma model. This model has two parameters, which control respectively the average level and slope of the base correlation curve.

The model is tractable to implement, with straightforward calibration. The model can price bespoke baskets and tranchelets, as well as exotic products such as CDO^2 and long-short CDOs. Risk management is provided with risks given to both every individual credit and the two skew parameters. This allows hedging of skew parameters across tranches and baskets.

Using the European approximation to implement the model does, in theory, remove its fully dynamic character. But the actual barrier implementation, which is more dynamic, produces very similar calibrated prices.

References

[1] D. Applebaum, *Levy Processes and Stochastic Calculus* (Cambridge University Press, 2004).
[2] D. Brigo, A. Pallavicini and R. Torresetti, Calibration of CDO tranches with the dynamical generalized-poisson loss model, working paper (2006).
[3] Creditflux newsletter, *Rated Equity Deals Put on Hold as Second Correlation Crisis Bites* (3 November 2006).
[4] J. Hull and A. White, Valuation of a CDO and an nth to default CDS without monte carlo simulation, *Journal of Derivatives* **12**(2) (2004) 8–23.
[5] M. Joshi and A. Stacey, Intensity Gamma: a new approach to pricing portfolio credit derivatives, preprint (2005).
[6] E. Luciano and W. Schoutens, A multivariate jump-driven financial asset model, preprint (2005).
[7] D. Madan, P. Carr and E. Chang, The variance gamma process and option pricing, *European Finance Review* **2**(1) (1998) 79–105.
[8] T. Moosbrucker, Pricing CDOs with correlated variance gamma distributions, colloquium paper, Centre for Financial Research, University of Cologne (2006).
[9] W. Press, S. Teukolsky, W. Vetterling and B. Flannery, *Numerical Recipes in C* (Cambridge University Press, 1988).
[10] W. Schoutens, Jumps in credit risk modelling, presentation, King's College London (2006).
[11] M. Winkel, Introduction to levy processes, graduate lecture, Department of Statistics, University of Oxford (2004).

CLUSTER-BASED EXTENSION OF THE GENERALIZED POISSON LOSS DYNAMICS AND CONSISTENCY WITH SINGLE NAMES*

DAMIANO BRIGO[†], ANDREA PALLAVICINI[‡]
and ROBERTO TORRESETTI[§]

Credit Models Banca IMI
Corso Matteotti 6, 20121 Milano, Italy
[†]*damiano.brigo@bancaimi.it*
[‡]*andrea.pallavicini@bancaimi.it*
[§]*roberto.torresetti@bancaimi.it*

We extend the common Poisson shock framework reviewed for example in Lindskog and McNeil [15] to a formulation avoiding repeated defaults, thus obtaining a model that can account consistently for single name default dynamics, cluster default dynamics and default counting process. This approach allows one to introduce significant dynamics, improving on the standard "bottom-up" approaches, and to achieve true consistency with single names, improving on most "top-down" loss models. Furthermore, the resulting GPCL model has important links with the previous GPL dynamical loss model in Brigo *et al.* [6], which we point out. Model extensions allowing for more articulated spread and recovery dynamics are hinted at. Calibration to both DJi-TRAXX and CDX index and tranche data across attachments and maturities shows that the GPCL model has the same calibration power as the GPL model while allowing for consistency with single names.

Keywords: Loss distribution; single name default dynamics; cluster default dynamics; calibration; generalized Poisson processes; stochastic intensity; spread dynamics; common Poisson shock models.

1. Introduction

The modeling of dependence or "correlation" between the default times of a pool of names is the key issue in pricing financial products depending in a nonlinear way on the pool loss. Typical examples are CDO tranches, forward start CDO's and tranche options.

A common way to introduce dependence in credit derivatives modeling, among other areas, is by means of copula functions. A copula corresponding to some preferred multivariate distribution is "pasted" on the exponential random variables

---

*This paper has been written partly as a response to criticism, suggestions, encouragements and objections to our earlier GPL paper posted on 12 January 2007 at SSRN.

triggering defaults of the pool names according to first jumps of Poisson or Cox processes. In general, if one tries to control dependence by specifying dependence across single default times, one is resorting to the so called "bottom-up" approach, and the copula approach is typically within this framework. Yet, such procedure cannot be extended in a simple way to a fully dynamical model in general. A direct alternative is to insert dependence among the default intensities dynamics of single names either by direct coupling between intensity processes or by introducing common factor dynamics. See for example the paper by Chapovsky et al. [8].

On the other side, one could give up completely single default modeling and focus on the pool loss and default counting processes, thus considering a dynamical model at the aggregate loss level, associated to the loss itself or to some suitably defined loss rates. This is the "top-down" approach pioneered by Bennani [2, 3] Giesecke and Goldberg [12], Sidenius, et al. [17], Schönbucher [16], Di Graziano and Rogers [9], Brigo, et al. [6], Errais et al. [11] among others. The first joint calibration results of a single model across indices, tranches attachments and maturities, available in Brigo et al. [6], show that even a relatively simple loss dynamics, like a capped generalized Poisson process, suffices to account for the loss distribution dynamical features embedded in market quotes. However, to justify the "down" in "top-down" one needs to show that from the aggregate loss model, possibly calibrated to index and tranche data, one can recover a posteriori consistency with single-name default processes. Errais et al. [11] advocate the use of random thinning techniques for their approach, but in general it is not clear whether a fully consistent single-name default formulation is possible given an aggregate model as the starting point. Interesting research on this issue is for example in Bielecki et al. [5], who play on markovianity of families of single name and multi-name processes with respect to different filtrations, introducing assumptions that limit the model complexity needed to ensure consistency.

Apart from these two general branches and their problems, mostly the above mentioned lack of dynamics in the classical "bottom-up" approach and the possible lack of "down" in the "top-down" approach, there is a special "bottom-up" approach that can lead to a loss dynamics resembling some of the "top-down" approaches above, and the model in Brigo et al. [6] in particular. This approach is based on the common Poisson shock (CPS) framework, reviewed in Lindskog and McNeil [15] with application in operational risk and credit risk for very large portfolios. This approach allows for more than one defaulting name in small time intervals, contrary to some of the above-mentioned "top-down" approaches.

The problem of the CPS framework is that it leads in general to repeated defaults. If one is willing to assume that single names and groups of names may default more than once, actually infinite times, the CPS framework allows one to model consistently single defaults and clusters defaults. Indeed, if we term "cluster" any (finite) subset of the (finite) pool of names, in the CPS framework different cluster defaults are controlled by independent Poisson processes. Starting from the clusters defaults one can easily go back either to single name defaults ("bottom-up") or to the default counting process ("top-down"). Thus we have a consistent framework for default counting processes and single name default, driven by independent clusters-default Poisson processes. In the "bottom-up" language, one sees that this approach leads to a Marshall–Olkin copula linking the first jump (default) times of single names. In the "top-down" language, this model looks very similar to the GPL model in Brigo *et al.* [6] when one does not limit the number of defaults.

In the credit derivatives literature the CPS framework has been used for example in Elouerkhaoui [10], see also references therein. Balakrishna [1] introduces a semi-analytical approach allowing again for more than one default in small time intervals and hints at its relationship with the CPS framework, showing also some interesting calibration results.

Troubles surface when one tries to get rid of the unrealistic "repeated default" feature. In past works it was argued that one just assumes cluster default intensities to be small, so that the probability that the Poisson process for one cluster jumps more than once is small. However, calibration results in Brigo *et al.* [6] lead to high enough intensities that make repeated defaults troublesome. The issue remains then if one is willing to use the CPS framework for dependence modeling in credit derivatives pricing and hedging.

In this paper we start from the standard CPS framework with repeated defaults and use it as an engine to build a new model for (correlated) single name defaults, clusters defaults and default counting process or portfolio loss. Indeed, if s is a set of names in the portfolio and $|s|$ is the number of names in the set s, we start from the (independent) cluster default Poisson processes $\widetilde{N}_s$ for example in Lindskog and McNeil [15] consistent with (correlated) single name k repeated default Poisson processes N_k, and build new default processes avoiding repeated single name and cluster defaults. We propose two ways to do this, the most interesting one leading to a new definition of cluster defaults $\widetilde{N}_s^2$ avoiding repetition and (correlated) single name defaults N_k^2 avoiding repetition as well, whose construction is detailed in Sec. 3.3. An alternative approach, based on an adjustment to avoid repeated defaults at single name level, and leading to (correlated) single name default processes N_k^1, is proposed in Sec. 3.2. This approach however leads to a less clear cluster dynamics in terms of the original cluster repeated default processes $\widetilde{N}_s$.

We then move on and examine the approach based on the non-repeated cluster and single name default processes $\widetilde{N}_s^2, N_k^2$, which we term "Generalized Poisson Cluster Loss model" (GPCL), detailing some homogeneity assumptions that can reduce the otherwise huge number of parameters in this approach. We calibrate the associated default counting process to a panel of index and tranche data across maturities, and compare the resulting model with the Generalized Poisson Loss (GPL) model in Brigo *et al.* [6]. The GPL model is similar to the GPCL model but lacks a clear interpretation in "bottom-up" terms, since we act on the default counting process, by capping it to the portfolio size, without any control of what happens either at single name or at clusters level. The GPCL instead allows us to understand what happens there. Calibration results are similar but now we may interpret the "top-down" loss dynamics associated to the default counting process in a "bottom-up" framework, however stylized this is, and have a clear interpretation of the process intensities also in terms of default clusters.

In Sec. 5 we hint at possible extensions leading to richer spread dynamics and recovery specifications. This, in principle, allows for more realism in valuation of products that depend strongly on the spread dynamics such as forward starting CDO tranches or tranche options. However, since we lack liquid market data for these products, we cannot proceed with a thorough analysis of the extensions. Indeed, the extensions are only presented as a proposal and to illustrate the fact that the model is easily generalizable. Further work is in order when data will become available.

2. Modeling Framework and the CPS Approach

We consider a portfolio of M names, typically $M = 125$, each with notional $1/M$ so that the total pool has unit notional. We denote with $\bar{L}_t$ the portfolio cumulated loss, with $\bar{C}_t$ the number of defaulted names up to time t ("default counting process") and we define $\bar{C}_t := C_t/M$ (default rate of the portfolio).

Since a portion of the amount lost due to each default is usually recovered, the loss is smaller than the default fraction. Thus,

$$0 \leq d\bar{L}_t \leq d\bar{C}_t \leq 1 \text{ for all } t, \quad \bar{L}_0 = 0, \bar{C}_0 = 0, \tag{2.1}$$

which in turn implies (but is *not* implied by) $0 \leq \bar{L}_t \leq \bar{C}_t \leq 1$.

Notice that with the notation dX_t, where X_t is a jump process which we assume to be right continuous with left limit, we actually mean the jump size of process X at time t if X jumps at t, and zero otherwise, or, in other terms, pathwise, $dX_t = X_t - X_{t^-}$, where in general we define $X_{t^-} := \lim_{h \downarrow 0} X_{t-h}$.

We can relate the cumulated loss process $\bar{L}_t$ and the re-scaled number of defaults $\bar{C}_t$ at any time t through the notion of *recovery rate at default* R_t,

$$d\bar{L}_t = d\bar{C}_t(1 - R_t), \quad \left(\bar{L}_t = \int_0^t (1 - R_u)d\bar{C}_u \right), \tag{2.2}$$

where R_t is a predictable process. We discuss technicalities and more advanced modeling proposals for R in Brigo et al. [7] and in Sec. 5 below.

The no-arbitrage condition (2.1) is met if R takes values in $[0, 1]$.

After these general considerations, we begin by briefly illustrating the common Poisson shock framework (CPS), reviewed for example in Lindskog and McNeil [15].

The occurrence of a default in a pool of names can be originated by different events, either idiosyncratic or systematic. In the CPS framework, the occurrence of the event number e, with $e = 1 \ldots m$, is modelled as a jump of a Poisson process $N^{(e)}$. Notice that each event can be triggered many times. Poisson processes driving different events are considered to be independent.

The CPS setup assumes unrealistically that a defaulted name k may default again. In the next section of the paper we try and limit the number of defaults of each name to one. For now, we assume that the r-th jump of $N^{(e)}$ triggers a default event for the name k with probability $p_{r,k}^{(e)}$, leading to the following dynamics for the single name default process N_k, defined as the process that jumps each time name k defaults:

$$N_k(t) := \sum_{e=1}^{m} \sum_{r=1}^{N^{(e)}(t)} I_{r,k}^{(e)}$$

where $I_{r,k}^{(e)}$ is a Bernoulli variable with probability $\mathbb{Q}\{I_{r,k}^{(e)} = 1\} = p_{r,k}^{(e)}$. Under the Poisson assumption for N^e and the Bernoulli assumption for $I_{r,j}^{(e)}$ it follows that N_k is itself a Poisson process. Notice however that the processes N_k and N_h followed by two different names k and h are not independent since their dynamics is explained by the same driving events.

The core of the CPS framework consists in mapping the single name default dynamics, consisting of the dependent Poisson processes N_k, into a multi-name dynamics explained in terms of *independent* Poisson processes $\widetilde{N}_s$, where s is a subset (or "cluster") of names of the pool, defined as follows.

$$\widetilde{N}_s(t) = \sum_{e=1}^{m} \sum_{r=1}^{N^{(e)}(t)} \sum_{s' \supseteq s} (-1)^{|s'|-|s|} \prod_{k' \in s'} I_{r,k'}^{(e)}$$

where $|s|$ is the number of names in the cluster s. In a summation, $s \ni k$ means we are adding up across all clusters s containing k, $k \in s$ means we are adding across all elements k of cluster s, while $|s| = j$ means we are adding across all clusters of size j and, finally, $s' \supseteq s$ means we are adding up across all clusters s' containing cluster s as a subset.

The non-trivial proof of the independence of $\widetilde{N}_s$ for different subsets s can be found in Lindskog and McNeil [15]. Notice that a jump in a $\widetilde{N}_s$ processes means that all the names in the subset s, *and only those names*, have defaulted at the jump time. We denote by $\widetilde{\lambda}_s$ the intensity of the Poisson process $\widetilde{N}_s(t)$, and we assume it to be deterministic for the time being, although we present extensions later.

Finally, we recall that the single name default dynamics in the CPS framework induces a Marshall–Olkin copula type dependence between the first jumps of the single name processes N_j. More precisely, if the random default times in the pool are modeled as the first jump times of the single name processes $N_1, \ldots, N_M$, then Lindskog and McNeil [15] show that the default times vector is distributed according to a multi-variate distribution whose survival copula is a M-dimensional Marshall–Olkin copula.

One does not need to remember the above construction. All that matters for the following developments are the independent clusters default Poisson processes $\widetilde{N}_s(t)$. These can be taken as fundamental variables from which (correlated) single name defaults and default counting processes follow. The single name dynamics can be derived based on these independent $\widetilde{N}_s$ processes in the so-called fatal shock representation of the CPS framework:

$$N_k(t) = \sum_{s \ni k} \widetilde{N}_s(t), \quad \text{or} \quad dN_k(t) = \sum_{s \ni k} d\widetilde{N}_s(t), \tag{2.3}$$

where the second equation is the same as the first one but in instantaneous jump form. We now introduce the process $Z_j(t)$, describing the occurrence of the simultaneous default of any j names whenever it jumps (with jump-size one):

$$Z_j(t) := \sum_{|s|=j} \widetilde{N}_s(t). \tag{2.4}$$

Notice that each $Z_j(t)$, being the sum of independent Poisson processes, is itself Poisson. Further, since the clusters corresponding to the different $Z_1, Z_2, \ldots, Z_M$ do not match, the $Z_j(t)$ are independent Poisson processes.

The multi-name dynamics, that is the default counting process Z_t for the whole pool, can be easily derived by carefully adding up all the single name contributions.

$$Z_t := \sum_{k=1}^{M} N_k(t) = \sum_{k=1}^{M} \sum_{s \ni k} \widetilde{N}_s(t) = \sum_{k=1}^{M} \sum_{j=1}^{M} \sum_{s \ni k, |s|=j} \widetilde{N}_s(t) = \sum_{j=1}^{M} j \sum_{|s|=j} \widetilde{N}_s(t),$$

leading to the relationship which links the set of dependent single name default processes N_k with the set of independent and Poisson distributed counting processes Z_j:

$$\sum_{k=1}^{M} N_k(t) = \sum_{j=1}^{M} j Z_j(t) =: Z_t \tag{2.5}$$

Hence, the CPS framework offers us a way to consistently model the single name processes along with the pool counting process taking into account the correlation structure of the pool, which remains specified within the definition of each cluster process $\widetilde{N}_s$. Notice, however, that the Z_t/M process is not properly the re-scaled number of defaults $\bar{C}_t$, since the former can increase without limit, while the latter is bounded in the $[0, 1]$ interval. We address this issue in Sec. 3 below, along with the issue of avoding repeated single names and cluster defaults.

Finally, notice that one way of looking at the Z_t process in the aggregate is the compound Poisson process, see Brigo et al. [7] for the details. We notice that also Di Graziano and Rogers [9] in some of their formulations obtain a compound Poisson process for the loss distribution.

3. Avoiding Repeated Defaults

In the above framework we have a fundamental problem, due to repeated jumps of the same Poisson processes. Indeed, if the jumps are to be intepreted as defaults, this leads the above framework to unrealistic consequences. Repeated defaults would occur both at the cluster level, in that a given cluster s of names may default more than once, as $\widetilde{N}_s$ keeps on jumping, and at the single name level, since each name

k keeps on defaulting as the related Poisson process N_k keeps on jumping. These repetitions would cause the default counting process Z_t to exceed the pool size M and to grow without limit in time.

There are two main strategies to solve this problem. Both take as starting points the cluster repeated-default processes $\widetilde{N}_s$ and then focus on different variables. They can be summarized as follows.

Strategy 1 (Single-name adjusted approach). Force single name defaults to jump only once and deduce clusters jumps consistently.

Strategy 2 (Cluster adjusted approach). Force clusters to jump only once and deduce single names defaults consistently.

The two choices have different implications, and we explore both of them in the following, although we anticipate the second solution is more promising.

If one gives up single names and clusters, and focuses only on the default counting process and the loss (throwing away the "bottom-up" interpretation), there is a third possible strategy to make the default counting process above consistent with the pool size:

Strategy 0 (Default-counting adjusted approach). Modify the aggregated pool default counting process so that this does not exceed the number of names in the pool.

Strategy 0 addresses the problem of the CPS framework at the default counting level. In the basic CPS framework, the link between the re-scaled pool counting process Z_t/M, which can increase without limit, and the re-scaled number of defaults $\bar{C}_t$, that must be bounded in the $[0, 1]$ interval, is not correct. This forbids in principle to model $\bar{C}_t$ as Z_t/M. In the CPS literature this problem is not considered usually. Lindskog and McNeil [15] for instance suppose that the default intensities of the names are so small to lead to negligible "second-default" probabilities. If this assumption were realistic, this would allow for adopting Z_t/M as a model for $\bar{C}_t$ and strategy 0 would not be needed. However, in our calibration results in Brigo *et al.* [6] we find that intensities are large enough to make repeated defaults unacceptable in practice.

3.1. *Default-counting adjustment: GPL model (Strategy 0)*

One possibility is to consider the pool counting process Z_t merely as a driving process of some sort for the market relevant quantities, namely the cumulated portfolio loss $\bar{L}_t$ and the re-scaled number of defaults $\bar{C}_t$. This candidate underlying process Z_t is non-decreasing and takes arbitrarily large values in time. The portfolio cumulated loss and the re-scaled number of defaults processes are non-decreasing, but limited to the interval $[0, 1]$. Thus, we may consider a deterministic non-decreasing function $\psi : \mathbb{N} \cup \{0\} \to [0, 1]$ and we define either the counting or loss process as $\psi(Z_t)$. In the first part of Brigo *et al.* [6] we go for the former choice, by capping the counting process coming from single name repeated defaults, assuming

$$\bar{C}_t := \psi_{\bar{L}}(Z_t) := \min(Z_t/M, 1), \tag{3.1}$$

where $M > 0$ is the number of names in the portfolio, while in the second part of Brigo et al. [6] we adopt the latter choice,

$$\bar{L}_t := \psi_{\bar{L}}(Z_t) := \min(Z_t/M', 1), \tag{3.2}$$

where $1/M'$, with $M' \geq M > 0$, is the minimum jump-size allowed for the loss process, leading to more refined granularity solutions. The quantity that is not modelled directly between $\bar{C}_t$ and $\bar{L}_t$ can be obtained from the one modelled directly through explicit assumptions on the recovery rate. We discuss recovery assumptions in general below, in Sec. 5.

This approach has the drawback of breaking the relationship (2.5) which links the single name processes N_k with the counting processes Z_j. We can still write the counting processes as a function of the repeated default counting process Z_t under formula (3.1):

$$Z_j^0(t) = \int_0^t 1_{\{dZ_u = j, Z_{u^-} \leq M - j\}} = \int_0^t 1_{\{Z_{u^-} \leq M - j\}} dZ_j(u),$$

but we have clearly no link with single names.

This can be considered a viable approach, if we are interested only in the collective dynamics of the pool without considering its constituents, i.e., in the aggregate loss picture typical of many "top-down" approaches.

3.2. Single-name adjusted approach (Strategy 1)

In order to avoid repeated defaults in single name dynamics, we can introduce constraints on the single name dynamics ensuring that each single name makes only one default. Such constraints can be implemented by modifying Eq. (2.3) in order to allow for one default only. Given the same repeated cluster processes $\tilde{N}_s$ as before, we *define* the new single name default processes N_k^1 replacing N_k as solutions of the following modification of Eq. (2.3) for the original N_k:

$$dN_k^1(t) := (1 - N_k^1(t^-)) \sum_{s \ni k} d\tilde{N}_s(t) \tag{3.3}$$

$$= \sum_{s \ni k} d\tilde{N}_s(t) \prod_{s \ni k} 1_{\{\tilde{N}_s(t^-)=0\}}$$

Interpretation: This equation amounts to say that name k jumps at a given time if some cluster s containing k jumps (i.e., $\tilde{N}_s$ jumps) and if no cluster containing name k has ever jumped in the past.

We can compute the new cluster defaults $\tilde{N}_s^1$ consistent with the single names N_k^1 as

$$d\tilde{N}_s^1(t) = \prod_{j \in s} dN_j^1(t) \prod_{j \in s^c} (1 - dN_j^1(t)) \tag{3.4}$$

where s^c is the set of all names that do not belong in s.

Now, we can use Eq. (2.5) with the N_k^1 replacing the N_k, to calculate how the new counting processes Z_j^1 are to be defined in terms of the new single names default dynamics:

$$\sum_{k=1}^{M} dN_k^1(t) = \sum_{k=1}^{M} (1 - N_k^1(t^-)) \sum_{s \ni k} d\tilde{N}_s(t) = \sum_{k=1}^{M} (1 - N_k^1(t^-)) \sum_{j=1}^{M} \sum_{s \ni k, |s|=j} d\tilde{N}_s(t)$$

$$= \sum_{j=1}^{M} \sum_{|s|=j} d\tilde{N}_s(t) \sum_{k \in s} (1 - N_k^1(t^-)) = \sum_{j=1}^{M} \sum_{|s|=j} d\tilde{N}_s(t) \sum_{k \in s} \prod_{s' \ni k} 1_{\{\tilde{N}_{s'}(t^-)=0\}}.$$

This expression should match $dZ^1(t) := \sum_j j\, dZ_j^1(t)$, so that the counting processes are to be defined as

$$dZ_j^1(t) := \frac{1}{j} \sum_{|s|=j} d\tilde{N}_s(t) \sum_{k \in s} \prod_{s' \ni k} 1_{\{\tilde{N}_{s'}(t^-)=0\}} \tag{3.5}$$

The intensities of the above processes can be directly calculated in terms of the density of the process compensator. We obtain by direct calculation

$$h_{N_k^1}(t) = \prod_{s \ni k} 1_{\{\tilde{N}_s(t^-)=0\}} \sum_{s \ni k} \tilde{\lambda}_s(t)$$

$$h_{Z_j^1}(t) = \frac{1}{j} \sum_{|s|=j} \tilde{\lambda}_s(t) \sum_{k \in s} \prod_{s' \ni k} 1_{\{\tilde{N}_{s'}(t^-)=0\}}$$

where in general we denote by $h_X(t)$ the compensator density of process X at time t, referred to as "intensity of X", and where $\tilde{\lambda}_s$ is the intensity of the Poisson process $\tilde{N}_s$.

Given exogenously the repeated Poisson "cluster" default building blocks $\tilde{N}_s$, the model $N_k^1, \tilde{N}_s^1, Z_j^1$ is a consistent way of simulating the single name processes, the cluster processes and the pool counting process from the point of view of avoiding repeated defaults. In particular, we obtain $\bar{C}_t := \sum_k N_k^1(t)/M = Z_t^1/M \leq 1$.

Notice, however, that the definition of N_k^1 in (3.3), even if it avoids repeated defaults of single names, is not consistent with the spirit of the original repeated cluster dynamics.

Consider indeed the following example.

Example. Consider two clusters $s = \{1,2,3\}$, $z = \{3,4,5,6\}$. Assume no name defaulted up to time t except for cluster z, in that in a single past instant preceding t names $3,4,5,6$ (and only these names) defaulted together ($\tilde{N}_z$ jumped at some past instant). Now suppose at time t cluster s jumps, i.e., names $1,2,3$ (and only these names) default, i.e., $\tilde{N}_s$ jumps for the first time.

Question: Does name 2 default at t?

According to our definition of N_2^1 the answer is yes, since no cluster containing name 2 has ever defaulted in the past. However, we have to be careful in interpreting what is happening at *cluster* level. Indeed, clusters z and s cannot both default since this way name 3 (that is in both clusters) would default twice. So we see that the actual clusters default of this approach, implicit in Eq. (3.4), do not have a clear intuitive link with repeated cluster defaults $\widetilde{N}_s$.

To simplify the parameters, we may assume the cluster intensities $\widetilde{\lambda}_s$ to depend only on the cluster size $|s| = j$. Then it is possible to directly calculate the intensity of the pool counting process $C = Z^1$ as

$$h_{Z^1}(t) = \left(1 - \frac{Z^1_{t-}}{M}\right) \sum_j j \binom{M}{j} \widetilde{\lambda}_j$$

where $\widetilde{\lambda}_j$ is the common intensity of clusters of size j.

We see that the pool counting process intensity h_{Z^1} is a linear function of the counting process $C = Z^1$ itself, as we can expect by general arguments for a pool of *independent* names (again with homogeneous intensities).

3.3. *GPCL model: Cluster-adjusted approach (Strategy 2)*

In the preceding sections we have seen that, if we are able to model all the repeated cluster defaults $\widetilde{N}_s$, we are able to describe the repeated default dynamics of both single names and the pool as a whole. Indeed, by knowing all the $\widetilde{N}_s$, we can directly compute the single name processes N_k and the aggregated counting processes Z_j by means of Eqs. (2.3) and (2.4).

In the previous section we have used the $\widetilde{N}_s$ exogenously as an engine to generate single name and aggregated defaults. This avoids repeated defaults of single names and a default rate exceeding 1, but is not consistent with the initial intuitive meaning of the $\widetilde{N}_s$'s as repeated clusters defaults.

The key to *consistently* avoid repeated cluster defaults (and subsequently single names) is to track, when a cluster jumps, which single-name defaults are triggered, and then force all the clusters containing such names not to jump any longer.

We may formalize these points by introducing the process $J_s(t)$ defined as

$$J_s(t) := \prod_{k \in s} \prod_{s' \ni k} 1_{\{\widetilde{N}_{s'}(t)=0\}} = \prod_{s': s' \cap s \neq \emptyset} 1 \widetilde{N}_{s'}(t) = 0$$

The process $J_s(t)$ is equal to 1 at starting time and it jumps to 0 whenever a cluster containing one or more elements of s jumps. Or one may view the process J_s as being one when none of the names in s have defaulted and 0 when some names in s have defaulted. Notice that $J_s(t) = 1$ implies $1_{\{\widetilde{N}_s(t)=0\}}$ but not vice versa.

We now correct the cluster dynamics by avoiding repeated clusters defaults. We define as new cluster dynamics the following:

$$d\widetilde{N}^2_s(t) = J_s(t^-)d\widetilde{N}_s(t). \tag{3.6}$$

Interpretation: Every time a repeated cluster default process $\widetilde{N}_s$ jumps, this is a jump in our "no-repeated-jumps" framework only if no name contained in s has defaulted in the past, i.e., if no cluster intersecting s has defaulted in the past.

Once the clusters defaults are given, single name defaults follow easily. We can change Eq. (2.3) and define the single name dynamics as

$$dN_k^2(t) := \sum_{s \ni k} d\widetilde{N}_s^2 = \sum_{s \ni k} J_s(t^-) d\widetilde{N}_s(t). \tag{3.7}$$

Now, we can use Eq. (2.4) to see how the counting processes Z_j are to be re-defined in terms of our new cluster dynamics (3.6). We obtain

$$dZ^2{}_j := \sum_{|s|=j} d\widetilde{N}_s^2 = \sum_{|s|=j} J_s(t^-) d\widetilde{N}_s(t). \tag{3.8}$$

The pool counting process reads

$$dZ^2 = \sum_{j=1}^{M} j \sum_{|s|=j} d\widetilde{N}_s^2 = \sum_{j=1}^{M} j \sum_{|s|=j} J_s(t^-) d\widetilde{N}_s(t). \tag{3.9}$$

If not for the cluster-related indicators $J_s(t^-)$, Z^2 would be a generalized Poisson process. That is why we term the model $N_k^2, \widetilde{N}_s^2, Z^2{}_j$ the Generalized Poisson Cluster-adjusted Loss model (GPCL).

Recall that we can always consider cluster dynamics as defined by single name dynamics rather than directly. That is, we can define

$$d\widetilde{N}_s^2(t) = \prod_{j \in s} dN_j^2(t) \prod_{j \in s^c} (1 - dN_j^2(t)) \tag{3.10}$$

This way the cluster s defaults, i.e., $\widetilde{N}_s^2$ jumps (at most once), when (and only when) all single names in cluster s jump at the same time (first product), provided that at that time no other name jumps (second product).

One can check that (3.10) and (3.6) are indeed consistent if the single name dynamics is defined by (3.7).

To appreciate how this second strategy formulation improves on the first strategy, we consider again our earlier example.

Example (Reprise). Consider the same example as in Sec. 3.2 up to the Question: "Does name 2 default at t?"

According to our definition of N_2^2 the answer is now NO, since the cluster $z = \{3, 4, 5, 6\}$, intersecting the s currently jumping (they both have name 3 as element), has already defaulted in the past. Thus we see a clear difference between strategies 1 and 2. With strategy 2 name 2 does not default when s jumps, with strategy 1 it does. Notice that strategy 2 is more consistent with the original spirit of the repeated cluster defaults $\widetilde{N}_s$. Indeed, if cluster $z = \{3, 4, 5, 6\}$ has defaulted in the past (meaning that $\widetilde{N}_z$ has jumped), $s = \{1, 2, 3\}$ should never be allowed to default, since it is impossible that now "exactly the names $1, 2, 3$ default", given that 3 has already defaulted in z.

The intensities of the above processes can be directly calculated as densities of the processes compensators. We obtain by direct calculation, given that $J_s(t)$ is known given the information (and in particular the $\widetilde{N}_s$) at time t:

$$h_{N_k^2}(t) = \sum_{s \ni k} J_s(t^-)\widetilde{\lambda}_s(t), \quad h_{Z_j^2}(t) = \sum_{|s|=j} J_s(t^-)\widetilde{\lambda}_s(t) \tag{3.11}$$

Remark 3.1 (Self-affecting features). Notice that in the GPCL model the single name intensities $h_{N_k^2}(t)$ are stochastic, since they depend on the process J_s. Moreover, the single name intensities are affected by the loss process. In particular, the intensity of a single-name jumps when one of the other names jumps. Consider for example a name k that has not defaulted by t, with intensity $h_{N_k^2}(t)$, and one path where there are no new defaults until $t' > t$, when name k' defaults. Now all clusters s containing k' have $J_s(t') = 0$ so that

$$h_{N_k^2}(t') = \sum_{s \ni k} J_s(t'^-)\widetilde{\lambda}_s(t') = \sum_{s \ni k} J_s(t^-)\widetilde{\lambda}_s(t') - \sum_{s \supseteq \{k,k'\}} J_s(t^-)\widetilde{\lambda}_s(t')$$

We see that the the kth name intensity reduces when k' defaults, and it reduces of the second summation in the last term.

At first sight this is a behaviour that is not ideally suited to intensities. For example, looking at the loss feedback present in the default intensities of Hawkes-processes (see [11] for Hawkes processes applied to default modeling), one sees that intensities are self-exciting, in that they *increase* when a default arrives. As soon as one name defaults, the intensities of the pool jump up, as is intuitive. However, Errais *et al.* [11] (but also [16] and others) assume there is only one default at a time. We are instead assuming there may be more than one default in a single instant. Therefore the self-exciting feature is somehow built in the fact that more than one name may default at the same instant. In other terms, instead of having the intensity of default of a related name jumping up of a large amount, implying that the name could default easily in the next instants, we have the two names defaulting together. From this point of view cluster defaults embed the self-exciting feature, although in an extreme way.

The best way to summarize our construction is through the three equations defining respectively cluster defaults, single name defaults and default counting processes:

$$d\widetilde{N}_s^2(t) = J_s(t^-)d\widetilde{N}_s(t), \quad \boxed{dN_k^2(t) := \sum_{s \ni k} d\widetilde{N}_s^2(t), \ dZ_j^2(t) := \sum_{|s|=j} d\widetilde{N}_s^2(t)}$$

Notice that once the new cluster default processes $\widetilde{N}_s^2$ are properly defined, single name and default counting processes follow immediately in what is indeed the only possible relationships that make sense for connecting clusters fatal shocks to single name defaults and to default counting processes. With our particular choice for the cluster defaults $\widetilde{N}_s^2$ dynamics we start from the repeated cluster defaults $\widetilde{N}_s$ dynamics and correct it to avoid repeated defaults at a cluster level. Then everything follows for default counting and single names.

Notice that the s-cluster intensity $J_s(t^-)\widetilde{\lambda}_s(t)$ strongly reminds us of what we do with Poisson (or more generally Cox) processes to model single name defaults. The default time τ_k of the single name k is modeled as the first jump of a Poisson process with intensity $\lambda_k(t)$, and then the process is killed after the first jump in order to avoid repeated defaults. This way the intensity of the default time τ_k is $1_{\{\tau_k > t\}}\lambda_k(t)$. What we do is similar for clusters: we start from clusters with repeated jumps $\widetilde{N}_s$ and then we kill the repeated jumps through an indicator $J_s(t)$, replacing the simpler indicator $1_{\{\tau_k > t\}}$ of the single-name case.

If, as before, *we assume the cluster intensities $\widetilde{\lambda}_s$ to depend only on the cluster size*, $\widetilde{\lambda}_s = \widetilde{\lambda}_{|s|}$, it is possible to directly calculate the intensity of the pool counting process $Z^2(t) := \sum_j j Z_j^2(t)$. We obtain

$$h_{Z^2}(t) = \sum_j j \binom{M - Z_{t^-}^2}{j} \widetilde{\lambda}_j$$

where $\widetilde{\lambda}_j$ is the common intensity of clusters of size j. The pool counting process intensity is a nonlinear function of the counting process, taking into account the co-dependence of single name defaults.

3.4. *Comparing models in a simplified scenario*

It is interesting to compare the relationships between the pool counting process C_t and its intensity across the different formulations we considered above.

Here, we summarize the approaches shown above *in the case cluster intensities depend only on the cluster size*, $\lambda_s = \lambda_{|s|}$.

1. **Repeated defaults.** The counting process can increase without limit, as implictly done in Lindskog and McNeil [15].

$$C_t = Z_t, \quad h_Z(t) = h_C(t) = \sum_{j=1}^{M} j \binom{M}{j} \widetilde{\lambda}_j(t)$$

2. **Strategy 0.** The counting process is bounded by the mapping $\psi(\cdot) := \min(\cdot, M)$, as in Brigo *et al.* [6]. This is the **Generalized Poisson Loss (GPL) model.**

$$C_t = \min(Z_t, M), \quad h_0(t) := h_C(t) = \sum_{j=1}^{M} \min(j, (M - Z_{t-})^+) \binom{M}{j} \widetilde{\lambda}_j(t)$$

3. **Strategy 1.** The counting process is bounded by forcing each single name to jump at most once. A dynamics, leading to a similar form of the intensity, is considered also in Elouerkhaoui [10].

$$C_t = Z_t^1, \quad h_1(t) := h_C(t) = \left(1 - \frac{Z_{t-}^1}{M}\right) \sum_{j=1}^{M} j \binom{M}{j} \widetilde{\lambda}_j(t)$$

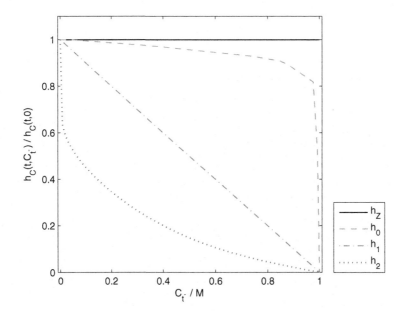

Fig. 1. The relationships between the pool counting process C_{t-}/M and its intensity ratio $h_C(t; C_{t-})/h_C(t; C_{t-} = 0)$ in the four different cases summarized in Sec. 3.4. The cluster intensities for the GPL and GPCL models are listed in the rightmost columns of the two panels of Table B.3.

4. Strategy 2. The counting process is bounded by forcing clusters dynamics to give raise to at most one jump in each single name. This is the **Generalized Poisson Cluster Loss (GPCL) model**.

$$C_t = Z_t^2, \quad h_2(t) := h_C(t) = \sum_{j=1}^{M} j \binom{M - Z_{t-}^2}{j} \widetilde{\lambda}_j(t)$$

In Fig. 1 we plot $h_C(t; C_{t-})/h_C(t; C_{t-} = 0)$ against C_{t-}/M in the four cases. The cluster intensities $\widetilde{\lambda}_j$ for the first and the third model are not relevant, since their influence cancels taking the ratio. The cluster intensities for the second and the fourth model are calibrated against the 10-year DJi-TRAXX tranche and index spreads on 2 October 2006 (see Table B.3).

Notice, further, that for any choice of the cluster intensities the pool intensities are monotonic non-increasing functions of the pool counting process, not explicitly depending on time.

4. The GPCL Model Calibration

In Brigo *et al.* [6] the GPL basic model $C_t = \min(Z_t, M)$ is calibrated to the index and its tranches for several maturities. Here we try instead the richer GPCL model $C_t = Z_t^2$ introduced above, allowing us in principle to model also cluster and single name defaults consistently. In the following we assume again that the cluster intensities $\widetilde{\lambda}_s$ depend only on the cluster size $|s|$. Moreover, as with the basic GPL model, we try calibration of multi-name products only, such as credit indices and CDO tranches, leaving aside single name data for the time being, in that we focus only on the improvement in calibration due to using a default counting process whose intensity has a clear interpretation in terms of default clusters. This will allow us, in further work, to include single names in the picture, since our GPCL framework allows us to do so explicitly.

The recovery rate is considered as a deterministic constant and set equal to $R = 40\%$. Thus, the underlying driving model definition is

$$C_t := Z^2(t) = \sum_{j=1}^{M} j\, Z_j^2(t), \quad \text{where} \quad dZ_j^2(t) \sim \text{Poisson}\left(\binom{M - Z_{t-}^2}{j}\widetilde{\lambda}_j(t)dt\right)$$

while the pool counting and loss processes are defined as

$$d\bar{C}_t := dZ_t^2/M, \quad d\bar{L}_t := (1 - R)\, dZ_t^2/M$$

Given our recovery assumption, the prices of the products to be calibrated, presented in the appendix, depend only on knowledge of the probability distribution of the pool counting process C_t. Thus, our main issue is to calculate this law as fast as possible. When dealing with dynamics derived from Poisson processes, there are different available calculation methods, depending on the structure of the intensities.

If the intensity does not depend on the process itself, or it does only in a simple way, then the probability distribution can be derived by means of Fast Fourier inversion of the characteristic function, when the latter is available in closed form. This method is described and used for the GPL model in Brigo *et al.* [6].

However, with the GPCL model, the dependence of the intensity of the pool counting process on the process itself prevents us either to calculate the relevant characteristic function in closed form or to use recursive techniques such as the Panjer method, described for example in Hess *et al.* [14].

Our choice then is to explicitly calculate the forward Kolmogorov equation satisfied by the probability distribution $p_{Z_t^2}(x) = \mathbb{Q}\{Z_t^2 = x\}$, namely

$$\frac{d}{dt}p_{Z_t^2}(x) = \sum_{y=0}^{M} A_t(x, y)p_{Z_t^2}(y)$$

where the transition rate matrix $A_t = (A_t(x, y))_{x,y=0,\dots,M}$ is given by

$$A_t(x, y) := \lim_{\Delta t \to 0} \frac{\mathbb{Q}\{Z_{t+\Delta t}^2 = x | Z_t^2 = y\}}{\Delta t} = \binom{M - y}{(x - y)}\widetilde{\lambda}_{(x-y)}(t)$$

for $x > y$,

$$A_t(y, y) := \lim_{\Delta t \to 0} \frac{\mathbb{Q}\{Z_{t+\Delta t}^2 = y | Z_t^2 = y\} - 1}{\Delta t} = -\sum_{j=1}^{M-y} \binom{M - y}{j}\widetilde{\lambda}_j(t)$$

for $x = y$, and zero for $x < y$.

In matrix form we write

$$\frac{d}{dt}\widehat{\pi}_t = A_t\widehat{\pi}_t, \quad \widehat{\pi}_t := \begin{bmatrix} p_{Z_t^2}(0) \ p_{Z_t^2}(1) \ p_{Z_t^2}(2) \ \ldots \ p_{Z_t^2}(M) \end{bmatrix}'$$

whose solution is obtained through the exponential matrix,

$$\widehat{\pi}_t = \exp\left(\int_0^t A_u du\right)\widehat{\pi}_0, \quad \widehat{\pi}_0 = [1\ 0\ 0...\ 0]'.$$

Matrix exponentiation can be quickly computed with the Padé approximation (see [13]), leading to a closed form solution for the probability distribution $p_{C_t} = \widehat{\pi}_t$ of the pool counting process C_t. This distribution can then be used in the calibration procedure. The integral of the matrix in the exponent will depend on the integrated cluster intensities

$$\widetilde{\Lambda}_j(t) = \int_0^t \widetilde{\lambda}_j(u)\,du$$

that are the actual calibration parameters. See Brigo *et al.* [7] for more details.

We assume the $\widetilde{\Lambda}_j$ to be piecewise linear in time, changing their values at payoff maturity dates. We have bM free calibration parameters, if we consider b maturities. Notice that many $\widetilde{\Lambda}_j(t)$ will be equal to zero for all maturities, meaning that we can ignore their corresponding counting process $Z_j^2(t)$. One can think of deleting all the modes with jump sizes having zero intensity and keep only the nonzero intensity ones. Call $\alpha_1 < \alpha_2 < \cdots < \alpha_n$ the jump sizes with nonzero intensity. Then one renumbers progressively the intensities according to the nonzero increasing α: Z_j^2 becomes the jump of a cluster of size α_j.

The calibration procedure for GPCL is implemented using the α_j in the same way as in Brigo *et al.* [6] for the GPL model. As concerns the GPCL intensities, in the tables we display $\binom{M}{\alpha_j}\widetilde{\Lambda}_j$, i.e., we multiply a cluster cumulated intensity for a given cluster size for the number of clusters with that size at time 0.

We also calibrate the GPL model, for comparison. In this paper we denote the GPL cumulated intensities for the α_j mode by Λ_j^0, which reads, using the link with repeated defaults, as $\Lambda_j^0 = \binom{M}{\alpha_j}\widetilde{\Lambda}_j$. Given the arbitrary *a-posteriori* capping procedure in Z, these $\widetilde{\Lambda}_j$ are not to be interpreted as cluster parameters, the only actual parameters being the Λ_j^0 directly, and they are to be interpreted as merely describing the pool counting process dynamic features.

More in detail, the optimal values for the amplitudes α_j in GPCL are selected, by adding non-zero amplitudes one by one, as follows, where typically $M = 125$:

1. set $\alpha_1 = 1$ and calibrate $\widetilde{\Lambda}_1$;
2. add the amplitude α_2 and find its best integer value by calibrating the cumulated intensities $\widetilde{\Lambda}_1$ and $\widetilde{\Lambda}_2$, starting from the previous value for $\widetilde{\Lambda}_1$ as a guess, for each value of α_2 in the range $[1, 125]$,

3. repeat the previous step for α_i with $i = 3$ and so on, by calibrating the cumulated intensities $\widetilde{\Lambda}_1, \ldots, \widetilde{\Lambda}_i$, starting from the previously found $\widetilde{\Lambda}_1, \ldots, \widetilde{\Lambda}_{i-1}$ as initial guess, until the calibration error is under a pre-fixed threshold or until the intensity $\widetilde{\Lambda}_i$ can be considered negligible.

The objective function f to be minimized in the calibration is the squared sum of the errors shown by the model to recover the tranche and index market quotes weighted by market bid-ask spreads:

$$f(\alpha, \widetilde{\Lambda}) = \sum_i \epsilon_i^2, \quad \epsilon_i = \frac{x_i(\alpha, \widetilde{\Lambda}) - x_i^{\text{Mid}}}{x_i^{\text{Bid}} - x_i^{\text{Ask}}} \tag{4.1}$$

where the x_i, with i running over the market quote set, are the index values S_0 for DJi-TRAXX index quotes, and either the index periodic premiums $S_0^{A,B}$ or the upfront premium rates $U^{A,B}$ for the DJi-TRAXX tranche quotes, see the Appendix for more details.

4.1. *Calibration results*

The calibration data set is the DJi-TRAXX main series on the run on 2 October 2006. In Table B.1 we list the CDO tranche spreads and the credit index spreads.

We calibrate three methods against such data set and we compare the results. They are listed in the tables.

1. The implied expected tranched loss method (hereafter ITL) described in Walker [19] or in Torresetti *et al.* [18]. It is a method which allows to check if arbitrage opportunities are present on the market by implying expected tranched losses satisfying basic no-arbitrage requirements.
2. The GPL model described in Brigo *et al.* [6] and summarized above, i.e., $C_t = \min(Z_t, M)$ with Z as in (2.5) (referred to before as strategy 0). Such model, due to the capping feature, is not compatible with any of the previously described single-name dynamics avoiding repeated defaults.
3. The GPCL model described in the present paper (strategy 2), which represents an articulated solution to the repeated defaults problem. We implement the simplified version with cluster intensity $\widetilde{\lambda}_s$ depending only on cluster size $|s|$.

First, we check that there are no arbitrage opportunities on 2 October 2006, by calibrating the ITL method. The calibration is almost exact and in Table B.5 we show the expected tranched losses implied by the method, which we can use as reference values when comparing the other two models.

Then we calibrate the GPL and GPCL models, and we obtain the calibration parameters presented in Table B.3, while the expected tranched losses implied by these two models are included in Table B.5. We point out that this is a joint calibration across tranche seniority and maturity, since we are calibrating all and every tranche and index quote with a single model specification. When looking at the outputs of the calibrated models on the different maturities, we see that both our models perform very well on maturities of 3 years, 5 years and 7 years, for which the calibration error is within the bid-ask spread. The 10 year maturity quotes are more difficult to recover, but both models are close to the market values, as we see from the left panel of Table B.6. Notice, however, that the GPCL model has a lower calibration error (10–20% better).

The probability distributions implied by the two dynamical models are similar at gross-grain view, as one can see in Fig. 2, but they differ if we observe the fine structure. Indeed, the tails of the two distributions show different bumps. The GPCL model shows a more complex pattern, and, as one can see from Table B.3, its highest mode is the maximum portfolio loss, while the GPL model has a less clear tail configuration.

We also apply the ITL, GPL and GPCL methods to the CDX index and tranches (see Table B.2 for market quotes), following the same procedure used for the

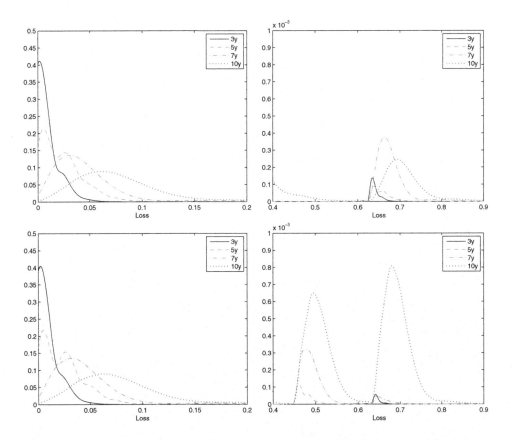

Fig. 2. Loss distribution evolution of GPL (upper panels) and of GPCL (lower panels) at all the quoted maturities up to ten years, drawn as a continuous line.

DJi-TRAXX above. We find better results, that are summarized in Table B.4 and in the right panel of Table B.6.

5. Extensions: Spread and Recovery Dynamics

In this final section we hint at possible extensions of the basic model to account for more sophisticated features.

The valuation of credit index forward contracts or options maturing at $\bar{t} = T_a$ requires the calculation of the index spread at those future times, which in turn depends on the default intensity evolution. Consider, for instance, the case of deterministic interest rates (or more generally interest rates independent of defaults) for an index whose default leg protects against losses in the index pool up to T_b and where the spread premium payments occur at $T_1, T_2, \ldots, T_b$. We have the spread expression at $\bar{t} < T_b$ as

$$
S_{\bar{t}} = \frac{\int_{\bar{t}}^{T_b} D(\bar{t}, t) \mathbb{E}_t[h_{\bar{L}}(t)] \, dt}{\sum_{i=1}^{b} \delta_i D(\bar{t}, T_i) \left(1 - \bar{C}_{\bar{t}} - \int_{\bar{t}}^{T_i} \mathbb{E}_{\bar{t}}[h_{\bar{C}}(t)] dt \right) 1_{\{T_i > \bar{t}\}}}
\tag{5.1}
$$

where $h_{\bar{L}}(t)$ is the default intensity of the cumulated portfolio loss process and $h_{\bar{C}}(t)$ is the default intensity of the re-scaled default counting process $\bar{C}$ (see for example Brigo *et al.* [6] or the Appendix, for a detailed description of credit index contracts) and $D(s,t)$ is the discount factor, often assumed to be deterministic, between times s and t.

The GPCL model presented in the previous sections has single-name and default counting intensities given by Eqs. (3.11). These intensities depend on which names have already defaulted. The dynamics of the index S_t (spread dynamics) can be enriched by more sophisticated modelling of the default intensities $h_{\bar{L}}(t)$ and $h_{\bar{C}}(t)$, by explicitly adding stochasticity to the Poisson intensities $\widetilde{\lambda}_j(t)$, e.g., resorting to the Gamma, scenario, CIR or Jump-CIR extensions of the model seen in Brigo *et al.* [7], to which we refer for further details.

Now we move to a discussion on the recovery. We introduced in (2.2) the notion of recovery at default R_t. In general, for ease of computation, we assume R_t to be a $\mathcal{G}_t$-adapted and left-continuous (and hence predictable) process taking values in the interval $[0, 1]$. On predictability of the recovery process see also Bielecki and Rutkowski [4]. Here $\mathcal{G}_t$ denotes the filtration consisting of default-free market information and of the default-count monitoring up to time t. Equation (2.2) leaves us with the freedom of defining only two processes among $\bar{L}_t$, $\bar{C}_t$ and R_t. The more natural approach would be modeling explicitly $(\bar{C}_t, R_t)$, obtaining $\bar{L}_t$, or modeling explicitly $(\bar{L}_t, R_t)$, obtaining $\bar{C}_t$, all of them adapted.

However, if we choose to model both $\bar{L}_t$ and $\bar{C}_t$ as $\mathcal{G}_t$-adapted processes and to infer R_t, we have to ensure that the resulting process R_t implicit in (2.2) is indeed left-continuous (and hence $\mathcal{G}_t$-predictable).

Indeed, in some formulations the predictability of the recovery is not possible. It is also a notion not always realistic: whether one or 125 names default in instant

$(t - dt, t]$ (i.e., $dC_t = 1$ or $dC_t = 125$, respectively), we would be imposing the recovery R_t to be the same in both cases and, in particular, to depend only on the information up to t^-.

However, under adapted-ness and left-continuity the recovery rate can be expressed also in terms of the intensities of the loss and default rate processes. From Eq. (2.2), by definition of compensator, we obtain

$$R_t = 1 - \frac{h_{\bar{L}}(t)}{h_{\bar{C}}(t)}. \tag{5.2}$$

Equation (5.2) shows that the recovery rate at default is directly related to the intensities of both the loss and the default rate processes. Thus, the choice for the intensity dynamics does induce a dynamics for the recovery rate.

If the recovery rate R is constant, as we assumed earlier in the paper, the portfolio total loss is forbidden, since bounded to be not greater than $1 - R$ on a unit portfolio notional.

We now briefly examine possible ways to model the loss more realistically, starting from a GPL or GPCL model formulated in terms of default counting process. This amounts to implicitly model the recovery rate, since the number of defaults and the loss are linked by the recovery at default. A first approach to implicitly model recovery rates consists in defining the cumulated portfolio loss $\bar{L}_t$ process as a deterministic function of the pool counting process $\bar{C}_t$ via a deterministic map. Set

$$\bar{L}_t := \psi(\bar{C}_t),$$

where ψ is a non-decreasing deterministic function with $\psi(0) = 0$ and $\psi(1) \leq 1$. What does this imply in terms of recovery dynamics? We can easily write

$$d\bar{L}_t = \sum_{k=1}^{M} \left[\frac{\psi(\bar{C}_{t-} + k/M) - \psi(\bar{C}_{t-})}{k/M} \right] 1_{\{d\bar{C}_t = k/M\}} d\bar{C}_t$$

which shows that the recovery at default in this case would not be predictable, depending explicitly from dC_t, except for very special ψ's.

A generalization based on a random process transformation (rather than a deterministic function) of the counting process leading to an implicit dynamics of the recovery process is presented in Brigo et al. [7], where the special case of a Gamma process is advocated.

6. Conclusions

We have extend the common Poisson shock (CPS) framework in two possible ways that avoid repeated defaults. The second way, more consistent with the original spirit of the CPS framework, leads to the Generalized-Poisson adjusted-Cluster-dynamics Loss model (GPCL). We have illustrated the relationship of the GPCL with our earlier Generalized Poisson Loss (GPL) model, pointing out that while the GPCL model shares the good calibration power of the GPL model, it further allows for consistency with single names.

Further research concerns recovery dynamics, calibration and analysis of forward start tranches and tranche options, when liquid quotes will be available, and analysis of calibration stability through history. A preliminary analysis of stability with the GPL model is however presented in Brigo *et al.* [6], showing good results. This is encouraging and leads to assuming the GPCL stability as well, although a rigorous check is in order in further work.

Acknowledgements

The authors are grateful to Aurélien Alfonsi, Marco Avellaneda, Norddine Bennani, Tomasz Bielecki, Giuseppe Castellacci, Charaf Chatbi, Dariusz Gatarek, Diego Di Grado, Youssef Elouerkhaoui, Kay Giesecke, Massimo Morini, Chris Rogers, Lutz Schlögl and James wood for helpful comments, criticism, further references and suggestions.

References

[1] B. S. Balakrishna, A semi-analytical parametric model for credit defaults, Working paper (2006) Available at http://www.defaultrisk.com/pp_crdrv128.htm

[2] N. Bennani, The forward loss model: A dynamic term structure approach for the pricing of portfolio credit derivatives, Working paper (2005). Available at http://defaultrisk.com/pp_crdrv_95.htm

[3] N. Bennani, A note on Markov functional loss models, Working Paper (2005) Available at http://www.defaultrisk.com/pp_cdo_01.htm

[4] T. Bielecki and M. Rutkowski, *Credit Risk: Modeling, Valuation and Hedging.* (Springer Verlag, Heidelberg, 2007).

[5] T. Bielecki, A. Vidozzi and L. Vidozzi, Pricing and hedging of basket default swaps and related derivatives, Preprint (2006).

[6] D. Brigo, A. Pallavicini and R. Torresetti, Calibration of CDO tranches with the dynamical Generalized-Poisson loss model, *Risk Magazine*, **May** (2007). Extended version available at http://www.defaultrisk.com/pp-crdrv117.htm

[7] D. Brigo, A. Pallavicini and R. Torresetti, Default correlation, cluster dynamics and single names: The GPCL dynamical loss model, (2007). Available at http://papers.ssrn.com/sol3/papers.cfm?abstract_id=956827

[8] A. Chapovsky, A. Rennie and P. A. C. Tavares, Stochastic Intensity Modelling for Structured Credit Exotics. Merrill Lynch working paper.

[9] G. Di Graziano and C. Rogers, A new approach to the modeling and pricing of correlation credit derivatives. Working paper available at www.statslab.cam.ac.uk/~chris/papers/cdo21.pdf

[10] Y. Elouerkhaoui, Pricing and hedging in a dynamic credit model, Citigroup Working paper, presented at the conference "Credit Correlation: Life After Copulas" (2006).

[11] E. Errais, K. Giesecke and L. Goldberg, Pricing credit from the top down with affine point processes. Working paper available at http://www.stanford.edu/dept/MSandE/people/faculty/giesecke/indexes.pdf

[12] K. Giesecke and L. Goldberg (2005). A top down approach to multi-name credit. Working paper available at http://www.stanford.edu/dept/MSandE/people/faculty/giesecke/topdown.pdf

[13] H. Golub and C. Van Loan (1983). *Matrix Computation*, (Johns Hopkins University Press, 1983).

[14] K. Hess, A. Liewald, K. Schmidt. An extension of Panjer's recursion, *Astin Bulletin* **32** (2002) 283–297.

[15] F. Lindskog and A. McNeil. Common Poisson shock models: applications to insurance and credit risk modelling. *Astin Bulletin* **33** (2003) 209–238.

[16] P. Schönbucher, (2005). Portfolio losses and the term structure of loss transition rates: a new methodology for the pricing of portfolio credit derivatives. Working paper available at http://defaultrisk.com/pp_model_74.htm

[17] J. Sidenius, V. Piterbarg, L. Andersen (2005). A new framework for dynamic credit portfolio loss modeling. http://defaultrisk.com/pp_model_83.htm

[18] R. Torresetti, D. Brigo and A. Pallavicini (2006). Implied Expected Tranched Loss Surface from CDO Data. http://www.damianobrigo.it/impliedetl.pdf

[19] M. Walker, (2006). CDO models. Towards the next generation: incomplete markets and term structure. http://defaultrisk.com/pp_crdrv109.htm

Appendix A. Market Quotes

The most liquid multi-name credit instruments available in the market are credit indices and CDO tranches (e.g., DJi-TRAXX, CDX).

The index is given by a pool names $1, 2, \ldots, M$, typically $M = 125$, each with notional $1/M$ so that the total pool has unitary notional. The index default leg consists of protection payments corresponding to the defaulted names of the pool. Each time one or more names default the corresponding loss increment is paid to the protection buyer, until final maturity $T = T_b$ arrives or until all the names in the pool have defaulted.

In exchange for loss increase payments, a periodic premium with rate S is paid from the protection buyer to the protection seller, until final maturity T_b. This premium is computed on a notional that decreases each time a name in the pool defaults, and decreases of an amount corresponding to the notional of that name (without taking out the recovery).

The discounted payoff of the two legs of the index is given as follows:

$$\text{DEFLEG}(0) := \int_0^T D(0,t) d\bar{L}_t$$

$$\text{PREMIUMLEG}(0) := S_0 \sum_{i=1}^{b} D(0,T_i) \int_{T_{i-i}}^{T_i} (1 - \bar{C}_t) dt$$

where $D(s,t)$ is the (deterministic) discount factor between times s and t. The integral on the right hand side of the premium leg is the outstanding notional on which the premium is computed for the index. Often the premium leg integral involved in the outstanding notional is approximated so as to obtain

$$\text{PREMIUMLEG}(0) = S_0 \sum_{i=1}^{b} \delta_i D(0, T_i)(1 - \bar{C}_{T_i})$$

where $\delta_i = T_i - T_{i-1}$ is the year fraction.

Notice that, differently from what will happen below with the tranches, here the recovery is not considered when computing the outstanding notional, in that only the number of defaults matters.

The market quotes the value of S_0 that, for different maturities, balances the two legs. If one has a model for the loss and the number of defaults one may impose that the loss and number of defaults in the model, when plugged inside the two legs, lead to the same risk neutral expectation (and thus price) when the quoted S_0 is inside the premium leg.

CDO tranches are obtained by "tranching" the loss of the pool consituting the index between the points A and B, with $0 \leq A < B \leq 1$.

$$\bar{L}_t^{A,B} := \frac{1}{B - A} \left[(\bar{L}_t - A) 1_{\{A < \bar{L}_t \leq B\}} + (B - A) 1_{\{\bar{L}_t > B\}} \right]$$

Once enough names have defaulted and the loss has reached A, the count starts. Each time the loss increases the corresponding loss change re-scaled by the tranche thickness $B - A$ is paid to the protection buyer, until maturity arrives or until the total pool loss exceeds B, in which case the payments stop.

The discounted default leg payoff can then be written as

$$\text{DEFLEG}(0; A, B) := \int_0^T D(0, t) d\bar{L}_t^{A,B}$$

As usual, in exchange for the protection payments, a premium rate $S_0^{A,B}$, fixed at time $T_0 = 0$, is paid periodically, say at times $T_1, T_2, \ldots, T_b = T$. Part of the premium can be paid at time $T_0 = 0$ as an upfront $U_0^{A,B}$. The rate is paid on the "survived" average tranche notional. If we assume that the payments are made on the notional remaining at each payment date T_i, rather than on the average in $[T_{i-1}, T_i]$, the discounted payoff of the premium leg can be written as

$$\text{PREMIUMLEG}(0; A, B) := U_0^{A,B} + S_0^{A,B} \sum_{i=1}^{b} \delta_i D(0, T_i)(1 - \bar{L}_{T_i}^{A,B})$$

where $\delta_i = T_i - T_{i-1}$ is the year fraction.

The tranche value is computed taking the (risk-neutral) expectation (in $t = 0$) of the discounted payoff consisting on the difference between the default and premium legs above. This expectation is set to zero when inserting the market quoted $S_0^{A,B}$ in the premium leg, finding the model parameters implicit in the loss and recovery assumptions reproducing the market.

The tranches that are quoted on the market refer to standardized pools, standardized attachment-detachment points $A - B$ and standardized maturities T.

Actually, for the i-Traxx and CDX pools, the equity tranche ($A = 0, B = 3\%$) is quoted by means of the fair $U_0^{A,B}$, while assuming $S_0^{A,B} = 500bps$. All other tranches are quoted by means of the fair $S_0^{A,B}$, assuming no upfront fee ($U_0^{A,B} = 0$).

Appendix B. Calibration Inputs and Outputs

Table B.1. DJi-TRAXX index and tranche quotes in basis points on 2 October 2006, along with the bid-ask spreads. Index and tranches are quoted through the periodic premium, whereas the equity tranche is quoted as an upfront premium. See Appendix A.

Att-Det		Maturities			
		3y	5y	7y	10y
Index		18(0.5)	30(0.5)	40(0.5)	51(0.5)
Tranche	0–3	350(150)	1975(25)	3712(25)	4975(25)
	3–6	5.50(4.0)	75.00(1.0)	189.00(2.0)	474.00(4.0)
	6–9	2.25(3.0)	22.25(1.0)	54.25(1.5)	125.50(3.0)
	9–12		10.50(1.0)	26.75(1.5)	56.50(2.0)
	12–22		4.00(0.5)	9.00(1.0)	19.50(1.0)
	22–100		1.50(0.5)	2.85(0.5)	3.95(0.5)

Table B.2. Same as Table 3 but for the CDX pool.

Att-Det		Maturities			
		3y	5y	7y	10y
Index		24(0.5)	40(0.5)	49(0.5)	61(0.5)
Tranche	0–3	975(200)	3050(100)	4563(200)	5500(100)
	3–7	7.90(1.6)	102.00(6.1)	240.00(48.0)	535.00(21.4)
	7–10	1.20(0.2)	22.50(1.4)	53.00(10.6)	123.00(7.4)
	10–15	0.50(0.1)	10.25(0.6)	23.00(4.6)	59.00(3.5)
	15–30	0.20(0.1)	5.00(0.3)	7.20(1.4)	15.50(0.9)

Table B.3. DJi-TRAXX pool. Left side: cumulated intensities, integrated up to tranche maturities, of the basic GPL model. Each row j corresponds to a different Poisson component with jump amplitude α_j. Right side: cumulated cluster intensities, integrated up to tranche maturities, and multiplied by the number of clusters of the same size at time 0. Each row j corresponds to a different cluster size α_j. The amplitudes/cluster-sizes not listed have an intensity below 10^{-7}. The recovery rate is 40%.

α_j	$\Lambda_j^0(T)$				α_j	$\binom{M}{\alpha_j}\tilde{\Lambda}_j(T)$			
	3y	5y	7y	10y		3y	5y	7y	10y
1	0.778	1.318	3.320	4.261	1	0.882	1.234	3.223	3.661
3	0.128	0.536	0.581	1.566	3	0.128	0.615	0.682	1.963
15	0.000	0.004	0.024	0.024	15	0.001	0.002	0.023	0.023
19	0.000	0.007	0.011	0.028	19	0.000	0.009	0.016	0.043
32	0.000	0.000	0.000	0.007	57	0.000	0.000	0.002	0.007
79	0.000	0.000	0.003	0.003	80	0.000	0.000	0.000	0.010
120	0.000	0.002	0.003	0.008	125	0.001	0.005	0.042	0.042

Table B.4. Same as Table B.3 but for the CDX pool.

α_j	$\Lambda_j^0(T)$				α_j	$\binom{M}{\alpha_j}\widetilde{\Lambda}_j(T)$			
	3y	5y	7y	10y		3y	5y	7y	10y
1	1.132	3.043	4.247	7.166	1	0.063	0.552	3.100	6.661
2	0.189	0.189	0.812	1.625	2	0.804	1.531	1.531	2.076
6	0.011	0.091	0.091	0.091	3	0.020	0.195	0.195	0.195
18	0.000	0.006	0.028	0.028	17	0.000	0.010	0.037	0.087
23	0.000	0.004	0.005	0.032	32	0.000	0.003	0.009	0.032
32	0.000	0.000	0.000	0.009	110	0.000	0.000	0.000	0.010
124	0.000	0.003	0.005	0.010	125	0.000	0.011	0.054	0.054

Table B.5. Implied expected tranched loss for the ITL, GPL and GPCL models. Results refer to DJi-TRAXX market.

Models	Maturities	Tranches					
		0–3	3–6	6–9	9–12	12–22	22–100
ITL	3y	18.6%	0.2%	0.1%	0.0%	0.0%	0.0%
	5y	44.6%	4.2%	1.2%	0.6%	0.2%	0.1%
	7y	71.0%	14.5%	4.3%	2.1%	0.7%	0.2%
	10y	91.6%	49.2%	14.1%	6.4%	2.2%	0.4%
GPL	3y	18.6%	0.2%	0.1%	0.1%	0.0%	0.0%
	5y	44.5%	4.2%	1.2%	0.6%	0.2%	0.1%
	7y	70.8%	14.6%	4.3%	2.1%	0.7%	0.2%
	10y	91.2%	47.2%	14.6%	6.4%	2.2%	0.4%
GPCL	3y	18.7%	0.2%	0.1%	0.0%	0.0%	0.0%
	5y	44.7%	4.2%	1.2%	0.6%	0.2%	0.1%
	7y	70.9%	14.6%	4.3%	2.1%	0.7%	0.2%
	10y	91.2%	47.5%	14.5%	6.4%	2.2%	0.4%

Table B.6. Calibration errors calculated with the GPL and GPCL models with respect to the bid-ask spread (i.e. ϵ_i in [4.1]) for tranches quoted by the market for the ten year maturity (see Tables B.1 and B.2). The left panel refers to DJi-TRAXX market quotes, while the right panel refers to CDX market quotes. Calibration errors for the other maturities are within the bid-ask spread and therefore they are not reported. The recovery rate is 40% .

	Att-Det	DJi-TRAXX 10y				Att-Det	CDX 10y	
		GPL	GPCL				GPL	GPCL
Index		0.00	0.00		Index		0.00	−0.06
Tranche	0–3	0.76	0.62		Tranche	0–3	1.43	1.60
	3–6	−2.35	−1.93			3–7	−0.45	−0.22
	6–9	1.21	1.04			7–10	0.22	0.25
	9–12	−0.40	−0.36			10–15	−0.08	−0.12
	12–22	0.02	0.02			15–30	0.01	0.07
	22–100	0.00	0.00					

STOCHASTIC INTENSITY MODELING FOR STRUCTURED CREDIT EXOTICS

ALEXANDER CHAPOVSKY, ANDREW RENNIE
and PEDRO TAVARES*

Merrill Lynch International
2 King Edward Street, London EC1A 1HQ, United Kingdom
pedro_tavares@ml.com

We propose a class of credit models where we model default intensity as a jump-diffusion stochastic process. We demonstrate how this class of models can be specialised to value multi-asset derivatives such as CDO and CDO^2 in an efficient way. We also suggest how it can be adapted to the pricing of option on tranche and leverage tranche deals. We discuss how the model performs when calibrated to the market.

1. Introduction

In the last several years the market for credit derivatives experienced an explosive growth both in terms of volume and innovation. The market in standard tranche CDOs became liquid (ITX, CDX, and others) and provides possibilities for hedging correlation risk. At the same time new exotic products are traded over the counter. These can be split broadly in two categories: default derivatives with complex payoffs (bespoke tranches, CDO^2 and others) and derivatives with payoffs that depend on spread levels and mark-to-market (options on tranches, leveraged super-senior, credit CPPI, *etc*). While the former require modelling of the defaults of individual single name credits, the latter require dynamical modelling of defaults and spread levels. This growth produces a need for new models for valuing and managing the risk.

The Gaussian copula model, [11] became the industry standard for the valuation of CDO. Emergence of a skew market in CDO of standard portfolios then gave rise to a number of extensions of this model that attempt to account for the structure of observed prices (for example, [1, 5, 13, 18]). All these approaches are similar in that they model loss distribution of a basket of credits at a given time horizon starting from default probabilities of single names through some copula function. These

---

*Corresponding author.

models have had variable success in calibrating to observed prices, depending on the complexity and flexibility of the copula and some became popular. Their main advantage is that the properties of single name credits are explicit inputs, which allows one to model more complicated derivatives such as CDO^2. The main drawback is that explicit intensity dynamics is absent from these models, which makes it impossible to use these models for modelling of more exotic credit derivatives, such as option on tranches.

Recently several authors proposed models that directly model the dynamics of the loss distribution of a given portfolio [15, 17]. Forward loss distribution is an input into the model which makes it possible to calibrate exactly to the structure of observed prices by construction. The model has dynamics as well so it is possible to price options on tranches. The main drawback of this model is that single name information is not an explicit input into the model and that makes it difficult to price bespoke basket CDOs or certain types of exotic structures such as CDO^2.

Direct modelling of stochastic credit default intensities for the valuation of CDO transactions was proposed by Duffie and Garleanu [6]. The main advantage of this approach is that the dynamics of default intensities for each credit can be specified which allows one to deal with CDO, CDO^2 and options on tranches within a single model, at least in principle. For a long time the perception was that this class of models is too complex and requires the use of Monte-Carlo methods for their implementation making efficient calibration impossible and deterring practitioners from using them in industrial applications. This motivated us to develop a modeling framework under which the default intensities of each single-name are modeled individually, yet the framework is still simple enough to allow for efficient calculations suited for use by practitioners. Very recently similar approaches were proposed in the literature [8, 10, 14].

The remainder of the paper is organised as follows. In Sec. 2 we described the basic setup of the model. Then in Sec. 3 we discuss parametrisation and calibration of the model to the CDO tranche market. In Sec. 4 we extend the framework to the pricing of option-like exotic credit derivatives. Finally we conclude in Sec. 5.

2. Model Setup

2.1. *Motivation*

From practical point of view a good model should have some important features:

- Have a parametrization that is intuitive.
- Be formulated in terms of single-name credits. This allows one to account for portfolio dispersion, bespoke tranches, CDO^2 and so on in a natural way.
- Calibrate to the term structure of single-name survival probabilities easily, preferably by construction.

- Be formulated in terms of local dynamics. This allows the model, at least in principle, to price path-dependent contracts and contracts which depend non linearly on future mark-to-market such as options on tranches.
- Calibrate to standard tranches, preferably by construction. This means in particular that it should produce enough default correlation and be flexible enough to match correlation skew.
- Suggest a hedging strategy with respect to single-name intensities (i.e., spread leverages) and model parameters (correlation skew hedge).

The model we discuss in this paper, on one hand, has most of these features and, on the other hand, is comparable to standard copula models in terms of its computational complexity when pricing vanilla tranches.

2.2. *Single credit dynamics*

The basic modeling quantity in our model is stochastic intensity, λ_i, of a credit name, i. Intensity should be positive for all times

$$\lambda_i(t) > 0, \tag{2.1}$$

and should calibrate to the term structure of survival probabilities

$$\mathrm{E}\left[e^{-\int_0^t \lambda_i dt}\right] = p_i(t), \tag{2.2}$$

where $p(t)$ is implied survival probability of the credit. The following ansatz solves the single name calibration equation

$$\lambda_i(t) = \lambda_i' - \lambda_i^c(t) + \lambda_i^f(t), \tag{2.3}$$

where $\lambda_i'(t)$ is an auxiliary random process. Compensator, $\lambda_i^c(t)$, and forward intensity, $\lambda_i^f(t)$, are deterministic functions of time which solve equations

$$\mathrm{E}\left[e^{-\int_0^t \lambda_i' dt}\right] = e^{-\int_0^t \lambda_i^c dt}, \quad e^{-\int_0^t \lambda_i^f dt} = p_i(t). \tag{2.4}$$

Term structure calibration and dynamics are separated in this ansatz: dynamics of λ_i' determines the compensator, λ_i^c, and survival probabilities, p_i, determine the forward intensities λ_i^f. However, the choice of dynamics is constrained by a requirement that total intensity stays positive for all times. In general, intensity dynamics, which satisfies (2.1), depends on the entire term structure of survival probabilities, $\lambda_i'[\lambda_i^f(t)]$, and is therefore different for different credits.

2.3. *Multiple credit dynamics*

The main object in modeling vanilla multi-name credit derivatives, such as CDO tranches, is the loss distribution of a basket of credits at various time horizons,

$P(t, L)$. This loss distribution depends in general on survival probabilities of underlying credits and on correlation structure of credit defaults. In our framework intensity realizations determine default probabilities

$$P(t, L) = \mathrm{E} \left[P(t, L | \lambda_i) \right], \tag{2.5}$$

where expectation is taken over all possible realizations of stochastic intensities, and $P(t, L | \lambda_i)$ is a loss distribution conditioned on realized trajectories of all intensities. Defaults are independent conditional on realized intensities. Therefore, $P(t, L | \lambda_i)$ is relatively easy to calculate, since $L = \sum l_i$, where l_i are independent single name losses with distributions determined by realizations of λ_i. In this setup, if dynamics of λ_i is known for every credit, one already can in principle perform all relevant calculations. This involves calculating outer expectation in the formula above by means of Monte-Carlo simulation inside which $P(t, L | \lambda_i)$ is calculated by recursion or another simulation. The outer simulation is of very high dimension and is not suitable for efficient calculations.

2.4. *Factorization of intensity dynamics*

One can try to reduce the dimensionality of the problem in order to make calculations more efficient. First of all, one can observe that conditional loss distribution of the basket, $P(L | \lambda_i)$, only depends on conditional survival probabilities of the credits, and therefore only on integrals $\int_0^t \lambda_i dt$, rather then full paths, λ_i. Second, one can try to reduce dimensionality further by considering some simple low dimensional factor model for dynamics of λ_i. However, as discussed above, the need to satisfy positivity constraint for intensity, (2.1), makes intensity dynamics dependent on the forward survival probability curve, $\lambda_i'[\lambda_i^f(t)]$, which is different for different names. This makes the dependence structure between the integrals of intensity and therefore survival probabilities, complicated in general. This fundamental interplay between the term structure of forward survival probabilities and factor dynamics for λ_i motivates the following parametrization for the joint dynamics

$$\int_0^t \lambda_i dt = \beta_i(t) \int_0^t y dt - \phi(t, \beta_i(t)) + \int_0^t \lambda_i^f dt,$$

$$\mathrm{E} \left[e^{-u \int_0^t y dt} \right] = e^{-\phi(t, u)}, \tag{2.6}$$

where y is the common random intensity driver, $\beta_i(t)$ and $\phi(t, u)$ are deterministic functions of time. Functions $\beta_i(t)$ should be chosen to guarantee positivity of intensity, (2.1), and are therefore credit specific, as discussed above. Function $\phi^c(t, u)$ is simply a characteristic exponent of variable $X(t)$,

$$X(t) = \int_0^t y dt, \quad \mathrm{E} \left[e^{-uX(t)} \right] = e^{-\phi(t, u)}. \tag{2.7}$$

To summarize, we have found a model specification, in which all conditional survival probabilities are expressed in terms of a single one dimensional variable, X,

$$p_i(t | X) = e^{-\int_0^t \lambda_i dt} = p_i(t) e^{-\beta_i(t) X + \phi(t, \beta_i(t))}, \tag{2.8}$$

in such a way that the term structure of implied survival probabilities, (2.2), is reproduced by virtue of the definition of $\phi(t, \beta_i(t))$. The positivity constraint (2.1) is satisfied by a credit specific choice of the functions $\beta_i(t)$, which in general depend of dynamics of X (induced by y), and term structure of the $p_i(t)$.

Expression (2.5) for the conditional loss distribution now reduces to

$$P(t, L) = \mathrm{E}\left[P(t, L|X)\right] = \int P(t, L|X)dP(X). \tag{2.9}$$

This expression demonstrates that the model has an equivalent copula formulation given by (2.8) and (2.9), with X playing the role of the central shock. The entire architecture practitioners have developed for copula models can then be employed here as well.

2.5. *Note on credit correlation*

The main goal of modeling the credit multi-name derivatives is to devise a mechanism that induces default correlation between different credit names. Here we would like to discuss briefly the alternatives. Defaults are modeled as a first jump of a Poisson process with some intensity and thus default of a name is correlated to its intensity by construction. Therefore, there are a number of options to induce correlations between the default events of two names:

- Default of a name causes changes in the intensity of other names;
- Default of a name causes defaults of other names directly (not affecting intensities);
- Intensity change of one name causes intensity changes of other names.

The first option corresponds to contagious default models[a] [16]. These models are attractive because they explicitly take into account the very intuitive effect of default of a name in the basket on other names in the basket. However, there is no reason why defaults of names outside of the basket should have no similar effect on the names in the basket. After all, choice of the basket is unlikely to affect real intensity dynamics of the single-names. Moreover, unless one models the entire credit universe simultaneously there are more names outside of the basket than inside. By neglecting effect of the names outside of the basket one ignores a potentially large effect.

The second option corresponds to models which allow several defaults at the same moment in time. Indeed, consider two names with correlated defaults within some time horizon. If one name defaults, intensity of other name should jump up accordingly because the other name is more likely to default now. If one insists on intensity not changing, the only way to allow non-zero default correlation is to

---

[a]Gaussian copula itself can be thought of as this type of model. Indeed, it is possible to come up with a dynamics (in which all intensities of the basket decrease if there are no defaults, and jump up if one of the names in the basket defaults), which reproduces loss distributions identical to that of Gaussian copula model.

allow for instantaneous defaults of both names. In other words, default correlation is local in time. Recently models which allow for multiple simultaneous defaults were discussed[b] [2, 4].

The third option, which we chose to follow, corresponds to the models where intensities of the credits have correlated dynamics and defaults are independent conditioned on the realization of intensity paths. Default correlations are indirectly induced by intensity correlations. It is well known that correlated intensity diffusion induces very small (parametrically small) default correlations, not nearly enough to explain levels of default correlations observed in the CDO market. One needs much more extreme dynamics to obtain reasonable default correlations and is therefore forced to add correlated jumps to intensity dynamics. As was noted already by Duffie and Garleanu [6], correlated jump-diffusion intensity dynamics induces much more default correlation. This model did not prove popular with the practitioners due to the necessity of relying on simulation methods for calculations and the availability of copula models in which semi-analytical calculations were possible. However, in this paper we discuss models with jump-diffusive intensity dynamics in which one can perform semi-analytical calculations, similar to those in copula models. More importantly, a new generation of credit derivatives (like options on tranches, LSS, credit CPPI, and others) cannot be modeled with copula models. Finally we would like to note that, at least on a finite time horizon, there is a similarity between these models and contagion models if shocks are originated by events inside as well as outside of the underlying basket. All of this makes stochastic intensity models attractive and serve as a motivation for this paper.

3. Model Parametrization and Calibration

In this section we describe possible model parameterizations and corresponding calibration results. The model is described by the dynamics of y and coefficients β_i. Together they determine the distribution $P(X)$ and compensator function $\phi(t, \beta_i)$. These are sufficient to calculate the loss distribution, $P(t, L)$, and therefore prices of CDO tranches.

Note that if the dynamics of y are such that $\phi(t, u)$ is known analytically, one knows not only all conditioned survival probabilities, but also the distribution of X, because they are related by

$$\mathrm{E}\left[e^{-uX(t)}\right] = \int e^{-uX(t)} dP(X) = e^{-\phi(t,u)}. \tag{3.1}$$

---

[b]Model suggested in [10] can also be viewed as belonging to this type of models. Business time of [10] used to calculate default probability experiences θ-like jumps. This is similar to intensity experiencing δ-like jumps in physical time. This means that correlation between defaults is localized in physical time.

$P(X)$ can then be found from $\phi(t, u)$ by inverse Laplace transform[c]. This makes it particularly attractive to look for models with analytical solutions for $\phi(t, u)$. Models with affine dynamics for y belong to this class and are extensively studied [3].

3.1. *Jump-only process*

As discussed previously, in order to induce enough default correlation, intensities should be subject to common jumps. We start therefore with a simple jump dynamics

$$dy = jdN, \tag{3.2}$$

where j is the randomly distributed jump size with probability distribution $p(j)$, and dN is a Poisson process with jump intensity Λ. This is an affine model, therefore $\phi(t, u)$ is explicitly known [3],

$$\phi(t, u) = \Lambda t \int dp(j) \left(1 - \frac{1 - e^{-jtu}}{jtu} \right) \tag{3.3}$$

We choose coefficients $\beta_i(t)$ to be equal to *average* intensity

$$\beta_i(t) = \bar{\lambda}_i(t), \quad \bar{\lambda}_i(t)t = \int_0^t \lambda_i^f dt. \tag{3.4}$$

With this choice, variable y and jumps j are dimensionless. We can also set, without loss of generality, $y(t = 0) = 0$.

Note that this model is already quite flexible. Indeed, one has the whole jump distribution, $p(j)$, as well as jump intensity, Λ, to calibrate CDO tranches. However, the requirement of positivity on intensity introduces bounds on $p(j)$ and Λ. In order to get some intuition about this consider flat term structure of intensity, $\lambda_i^f(t) = \lambda_i^f$. Then jump size, j, simply measures by how much intensity, λ_i, jumps relative to the forward intensity, λ_i^f. Also, in this case, $\beta_i(t) = \lambda_i^f$. Realized intensity is

$$\lambda_i(t) = \lambda_i^f y - \frac{\partial}{\partial t}\phi(t, \lambda_i^f) + \lambda_i^f, \tag{3.5}$$

where

$$\frac{\partial}{\partial t}\phi(t, u) = \Lambda \int dp(j) \left(1 - e^{-jtu} \right). \tag{3.6}$$

For $jtu \ll 1$ this function grows linearly, $\partial_t\phi(t, u) \sim t$, and then saturates exponentially at $jtu \gg 1$, $\partial_t\phi(t, u) \to \Lambda$. As y is always positive $\lambda_i^f - \partial_t\phi(t, \lambda_i^f)$ should be

---

[c]Sometimes inverse Laplace transformations are seen as difficult for numerical implementation due to slow convergence. There are, however, efficient algorithms, which resolve this issue. In particular we used algorithms described in [9].

positive to guarantee that $\lambda_i(t)$ is positive. This is guaranteed for at any time and for any jump size if

$$\Lambda < \lambda_i^f. \tag{3.7}$$

This is too restrictive in practice because common jump intensity, Λ, is constrained by the intensity of the tightest credit, which can be very small, resulting ultimately in very small default correlations. A more flexible bound is obtained if one constrains intensity to be positive only during the finite time, t, of the order of the maturity of the trade[d]. In that case it is enough if

$$jt\Lambda \ll 1 \tag{3.8}$$

as in this limit

$$\frac{\partial}{\partial t}\phi(t, u) = \Lambda tu \int dp(j)j. \tag{3.9}$$

Note that this bound does not depend on forward, λ_i^f. This bound is already quite flexible. For example, for a 10 year horizon it allows ten-fold intensity jumps with 0.1% intensity, or two-fold jumps with 5% intensity. Smaller, more frequent jumps can be outside of this bound. The main reason for the bound is the linear dynamics for intensity. The size of the jumps depends on the forward rather than on the current level of intensity. These small jumps, however, can be effectively described by diffusion, suggesting an extension to the model which we describe below.

Another problem might occur in presence of very large rare jumps. If intensity of these jumps is small enough no intensity bounds are violated, so there is no problem in principle. However, there might be a technical problem because very large jumps force one to calculate $P(X)$ for very large values of X, slowing down the calculation. However, very large jumps of intensity mean that intensity widens so much that names default very quickly. So an effective description of this behavior is to include a systemic default of all names with some intensity, λ_{sys}. One has to keep in mind that this is just an effective description of very large jumps, which is convenient to make calculations more efficient, but it is not needed in principle. An obvious constraint applies to λ_{sys}

$$\lambda_{\text{sys}} < \lambda_i^f. \tag{3.10}$$

Calibration to index tranche market values is shown in Tables 1 and 2. The model is able to fit the observed 5 year prices in both CDX and ITX markets very accurately. The results also exhibit three distinct jump scales: a small (<1) group of jumps, a second group of jumps of the order $1 < j < 10$ and finally a group of very large jumps, which for numerical reasons were replaced with a systemic component, as was described earlier. The fact that this structure arises naturally in calibrating the model is both interesting and reassuring because several authors have observed

---

[d]It is not unreasonable to do this because, in practice, intensity term structure is upwards sloping, relaxing the constraint for longer maturities

Table 1. Jump-only calibration of 5y DJIG on 20 September 2005 ($\Lambda = 100\%$).

Attach	Detach	Bid	Offer	Mid	Model	Jump	Prob
15%	30%	0.07%	0.08%	0.07%	0.07%	0.27	0.212
10%	15%	0.14%	0.16%	0.15%	0.14%	0.50	0.085
7%	10%	0.27%	0.3%	0.29%	0.29%	5.50	0.005
3%	7%	1.21%	1.24%	1.23%	1.21%	10.00	0.001
0%	3%	42.38%	42.88%	42.63%	42.39%	Syst.	0.0007

Table 2. Jump-only calibration of 5y ITX on 20 September 2005 ($\Lambda = 100\%$).

Attach	Detach	Bid	Offer	Mid	Model	Jump	Prob
12%	22%	0.06%	0.07%	0.06%	0.06%	0.27	0.410
9%	12%	0.1%	0.13%	0.11%	0.13%	0.55	0.399
6%	9%	0.21%	0.26%	0.24%	0.21%	8.00	0.003
3%	6%	0.76%	0.8%	0.78%	0.79%	10.03	0.001
0%	3%	24.5%	25.5%	25.%	24.55%	Syst.	0.0005

(see [2, 4, 18]) that models with 3-dimensional parameter structures are successful at recovering market prices.

3.2. *Jump-CIR process*

In calibrating the jump-only process we observed that jumps fall into groups according to their size. In particular we noted that a group of small jumps ($j < 1$) arises in the calibration. This *fine structure* is similar to diffusion so it is natural to describe small frequent jumps by adding a nonlinear diffusion to the dynamics of y:

$$y = y_D + y_J, \quad dy_D = \theta(\eta - y_D)dt + \sigma\sqrt{y_D}dW, \quad dy_J = jdN. \tag{3.11}$$

Here, affine diffusive dynamics is chosen to preserve tractability of $\phi(t, u)$, [3, 6].

We again show two example of calibration in Tables 3 and 4 but now focus on comparing 5 and 10 year trades to highlight certain features. The two calibrations are obtained separately with time-constant parameters but we ensured that all parameters except the jump intensity is held constant. While it is possible to obtain a more accurate calibration if we free the parameters from this restriction, the purpose here is to ensure as consistent a parametrisation across market prices as

Table 3. Jump-CIR calibration of 5y ITX on 3 November 2005 ($\theta = \eta = 1$, $\sigma = 35\%$, $\Lambda = 3\%$).

Attach	Detach	Bid	Offer	Mid	Model	Jump	Prob
12%	22%	0.05%	0.07%	0.06%	0.06%	2.10	0.65
9%	12%	0.1%	0.13%	0.12%	0.11%	7.00	0.1
6%	9%	0.21%	0.24%	0.23%	0.22%	Syst.	0.00053
3%	6%	0.91%	0.94%	0.93%	0.91%		
0%	3%	29.%	29.5%	29.25%	29.27%		

Table 4. Jump-CIR calibration of 10y ITX on 3 November 2005 ($\theta = \eta = 1$, $\sigma = 35\%$, $\Lambda = 1\%$).

Attach	Detach	Bid	Offer	Mid	Model	Jump	Prob
12%	22%	0.2%	0.25%	0.23%	0.25%	2.10	0.65
9%	12%	0.46%	0.56%	0.51%	0.45%	7.00	0.1
6%	9%	0.96%	1.06%	1.01%	1.%	Syst.	0.0019
3%	6%	5.05%	5.35%	5.2%	5.09%		
0%	3%	55.75%	57.25%	56.5%	56.09%		

possible, especially with respect to the term structure properties of the model. We concluded this could be achieved by changing only the jump intensity and systemic intensity. Given these considerations we observe that calibration is accurate but between the 5 and 10 year maturity we see a very sharp drop in jump intensity. This fact suggests the next generalization: to add mean reversion to the jump process to improve the term structure properties of the model.

3.3. *Non-linear jump-diffusion process*

Alternative specification of the model, which includes mean reversion of jumps, with analytic solution for $\phi(t, u)$ is canonic affine parametrization like in [6, 8, 14]:

$$dy = \theta(\eta - y)dt + \sigma\sqrt{y}dW + jdN, \tag{3.12}$$

where jumps are exponentially distributed and so $p(j)$ is parametrised by just one parameter. This choice provides less flexibility in calibrating to CDO tranches and so we generalized it further to account for any distribution of j. The cost of such extension is that the compensator is no longer analytic and we can no longer apply the inverse Laplace method. Instead the distribution $P(X)$ is determined by solving for

$$dy(t) = \mu(t, y)dt + \sigma(t, y)dW + jdN \tag{3.13}$$

$$dX(t) = y(t)\,dt. \tag{3.14}$$

where μ and σ are generic functions of time and process y. These can be defined by (3.12) but that is not strictly necessary. This is a very rich parametrization which can feature volatility smile, mean reversion and term structure of all parameters. As a result the calculation is more complex. A two-dimensional PIDE needs to be solved to determine $P(X)$. Once $P(X)$ is known, the compensator can be determined through (3.1).

Calibration results are shown in Tables 5 and 6. As before we aimed to calibrate the model with minimal term structure. In particular we keep all parameters constant expect jump and systemic intensities. We set $\mu(t, y) \equiv \mu(y) = \theta(y)(\eta - y)$ and $\sigma(t, y) \equiv \sigma(y)$. On average $\bar{\theta} \sim 1$ and $\bar{\sigma} \sim 30\%$. Spectrum of jumps remains constant in time. As expected the mean reversion effect on jumps reduces the term structure of calibrated jump intensity. It is also noticeable that calibration quality is worse than in earlier cases. Partially this is due to the fact that we sacrificed quality of fit

Table 5. Non-linear calibration of 5y ITX on 3 November 2005 ($\eta = 1$, $\bar{\theta} \sim$ 1, $\bar{\sigma} \sim 30\%$, $\Lambda = 1.00\%$).

Attach	Detach	Bid	Offer	Mid	Model	Jump	Prob
12%	22%	0.05%	0.07%	0.06%	0.05%	1.02	0.25
9%	12%	0.1%	0.13%	0.12%	0.11%	3.15	3.50
6%	9%	0.21%	0.24%	0.23%	0.24%	4.30	1.30
3%	6%	0.91%	0.94%	0.93%	0.9%	6.49	0.8
0%	3%	29.%	29.5%	29.25%	29.27%	Syst.	0.00053

Table 6. Non-linear calibration of 10y ITX on 3 November 2005 ($\eta = 1$, $\bar{\theta} \sim$ 1, $\bar{\sigma} \sim 30\%$, $\Lambda = 0.75\%$).

Attach	Detach	Bid	Offer	Mid	Model	Jump	Prob
12%	22%	0.2%	0.25%	0.23%	0.26%	1.02	0.25
9%	12%	0.46%	0.56%	0.51%	0.45%	3.15	3.50
6%	9%	0.96%	1.06%	1.01%	1.1%	4.30	1.30
3%	6%	5.05%	5.35%	5.2%	4.91%	6.49	0.8
0%	3%	55.75%	57.25%	56.5%	57.92%	Syst.	0.0019

for the sake of flatter term structure. Allowing for a steeper term structure, including spectrum of jumps will improve the calibration quality. In addition the model dynamics is now highly non-linear and therefore calibration procedure itself is now more subtle. More advanced methods of calibration are appropriate in this case.

3.4. *Idiosyncratic intensity dynamics*

Discussion so far was based on one factor model for intensity dynamics. That means that the intensity of all credits changes collectively. One might find this absence of idiosyncratic dynamics unnatural. It is easy to extend the model to allow for idiosyncratic dynamics by introducing credit specific factors. Schematically

$$\lambda_i = (\beta_i y - y^c) + (y_i - y_i^c) + \lambda_i^f, \tag{3.15}$$

where $\lambda_i^f(t)$ is forward, and y and y_i are common and idiosyncratic drivers respectively with their compensators

$$\mathrm{E}\left[e^{-\beta_i \int_0^t y \, dt}\right] = e^{-\int_0^t y^c dt}, \quad \mathrm{E}\left[e^{-\int_0^t y_i dt}\right] = e^{-\int_0^t y_i^c dt}. \tag{3.16}$$

This is a much more general specification then we considered previously and calculations are more complicated in this case. However, it is natural to think that prices of CDO tranches should be determined by collective rather then idiosyncratic dynamics. To see how this intuition manifests itself in our setup let us consider loss distribution of a credit basket

$$P(t, L) = \mathrm{E}\left[P(t, L | y, y_i)\right] = \mathrm{E}\left[\mathrm{E}\left[P(t, L | y, y_i)\right] | y\right], \tag{3.17}$$

where outer expectation is over y, and inner expectation is over y_i conditioned on y. Observe that, due to conditional independence of defaults, all terms in $P(t, L|y, y_i)$ are proportional to either conditional survival or conditional default probability of credit i. Therefore $P(t, L|y, y_i)$ is linear in $\exp(-\int_0^t (y_i - y_i^c)dt)$. By definition of y_i^c expectation of this quantity is 1, which means that loss distribution, $P(t, L)$, and CDO tranche prices do not depend on idiosyncratic dynamics. This argument is not completely correct because it neglects indirect effect through positivity of intensity constraint. Through this constraint idiosyncratic dynamics can have an effect on allowed values of credit's coupling to common driver β_i, and thus indirectly affect CDO tranche prices. If one assumes that constraint is satisfied, idiosyncratic effect completely drops out.

4. Application to Structured Credit Exotics

As discussed above, the stochastic intensity model described in this paper induces default copula and therefore allows one to model credit derivatives with payoffs dependent on the defaults of underlying credits (like, for example, CDO or CDO^2 tranches) by semi-analytic calculations or conditioned Monte-Carlo similar to conventional copula models. The model is also formulated in terms of the local dynamics of intensity of each name and therefore allows modelling of credit derivatives whose payoffs depend on the paths of loss and intensity, like CPPI on credit index, by direct simulation in the same self consistent model. In this section we consider the pricing of other types of exotic credit derivatives whose value depend in a nonlinear way on the mark-to-market value of other underlying credit derivatives. Derivatives of this type are options on CDO tranches, leveraged super-senior tranches with various triggers (loss, intensity and mark-to-market triggers), CDO tranches with counterparty risk, where counterparty credit risk is correlated with the credit risk of the underlying names, and so on. It is desirable to price these exotic derivatives in the same model as vanilla derivatives in order to achieve consistency and avoid arbitrage.

4.1. *Approximating model dynamics*

The main difficulty in modelling products of this type is a combination of very high dimension and a need to calculate the value of underlying derivatives potentially at every time step and for every realized value of market variables. One method to deal with this type of problems is the often called *American Monte-Carlo* [12], which relies on trying to approximately estimate the values of a relevant contract as a function of smaller number of variables while running the simulation. Here we try to approach the problem differently. We try to find a low dimensional model, which approximates the original high dimensional model. We then solve the pricing problem in a low dimensional approximating model by backward induction methods.

Consider a derivative with a large basket of credits as underlying. The price of this derivative at time t depends on realized defaults and realized term structure of intensities of the underlying credits at that time

$$V_t = V_t\left(I_t^{(i)}, \lambda_t^{(i)}\right), \tag{4.1}$$

where $I_t^{(i)}$ are default indicators and the $\lambda_t^{(i)}$ denote the implied term structure of survival probabilities. It satisfies equation

$$\frac{1}{B_t} V_t = \frac{1}{B_{t+\delta t}} \mathrm{E}\left[F(V_{t+\delta t})|I_t^{(i)}, \lambda_t^{(i)}\right] \tag{4.2}$$

where F is some nonlinear function. This is a backward induction equation of very high dimension. We will try to approximate this pricing problem by a low dimensional pricing problem which is more tractable. In doing so it is crucial to find factors which approximate the problem well. As discussed in previous section, in our framework current intensities of all credits in the basket are determined by a single common driver, y, which follows jump diffusion. We assume that the loss distributions of the basket, $P(t, L)$, are calculated using the appropriate methods of Secs. 2 and 3. In particular, loss distribution has representation

$$P(t, L) = \mathrm{E}\left[P(t, L|X)\right] = \int P(t, L|X)dP(X), \tag{4.3}$$

where variable X,

$$X_t = \int_0^t ydt, \tag{4.4}$$

is connected to the realized survival probability up to time t, and $P(t, L|X)$ is the loss distribution at time t conditioned on realized survival probabilities.

What should one choose as a minimal set of factors for the low dimensional approximating model? Any derivative on the credit basket must depend at least on the current level of intensities and realized loss. That is why y, corresponding to current intensities, and L, corresponding to realized loss, must be included in the minimal set of factors. Additionally, it is important to require that the approximating pricing model reproduces the loss distributions of the basket (and therefore all tranche prices), as implied by the full model, by construction. In order to achieve that we have to include X, corresponding to realized survival probabilities, in the minimal set of factors for our approximating model. In this way we will reproduce also conditioned loss distributions, $P(t, L|X)$, which can serve as the motivation to include X as a factor in the approximating model in its own right. Loss distribution then depends on realized survival probabilities, and thus on X, which in turn is correlated with the current level of intensity, y.

To summarize, our approximating model has three factors: y, L and X, with some associated dynamics induced by the full model, which we discuss below. Note, that information about individual credits does not feature in the approximated model explicitly. Instead, it enters implicitly through the loss distributions, which model dynamics has to reproduce. This makes our approximating model similar, in spirit, to dynamical loss models of [15, 17] in that it is trying to model loss of the basket as a dynamic variable.

In this low dimensional model the price of derivatives depends on the three factors discussed above

$$V_t = V_t(y_t, X_t, L_t), \tag{4.5}$$

and satisfies the following tree dimensional backward induction equation

$$\frac{1}{B_t} V_t = \frac{1}{B_{t+\delta t}} \mathrm{E} \left[F(V_{t+\delta t}) | y_t, X_t, L_t \right]. \tag{4.6}$$

Note that the price of derivatives in this model depends on variable X, which is connected to the realized survival probabilities. This may seem unnatural and even incorrect because intuitively the price should depend on current and future credit intensities, not past intensities. This seeming paradox is resolved as follows. Recall that we needed to include X as one of factors in order to reproduce loss distributions. These carry information about dispersion of credits in the basket. Imagine now that by the time t some assets defaulted. If credits in the basket are different, the value of the derivative depends on identities of defaulted credits. Imagine now that the only information available is the total loss up to the time t. To price the derivative one now has to assess which credits are more likely to have defaulted and which credits are more likely to remain in the basket. In order to do so one needs to know the past realized intensities, and therefore, the price of the derivative does indeed depend on X. In other words, X effectively captures dispersion of the credits in the basket.

To complete the specification of the model one needs to specify the dynamics of the drivers y, X and L. Dynamics of y is given by the same jump-diffusion process as that in the full model, (3.13),

$$dy_t = \mu_t(y)dt + \sigma_t(y)dW + j_t(y)dN, \tag{4.7}$$

where we changed notation slightly in order to make it consistent with this section. Note, that μ, σ can be functions of t and y, and j is a random variable with probability distribution, $p(j)$, which also can depend on t and y. Like in the full model, variable X follows

$$dX = ydt. \tag{4.8}$$

Finally loss, L, is a dynamical variable, with dynamics chosen to calibrate conditional loss distributions $P(t, L|X)$ for all t and X. We will find this transition probability neglecting, here and in the remainder of this Section, terms of higher orders in δt. Transition probability satisfies

$$P(t + \delta t, L_2|X_2) = \int dL_1 P(t + \delta t, L_2|t, L_1, X_1, y_1)P(t, L_1|X_1), \tag{4.9}$$

where X_1, L_1 and y_1 are state variables at time t, X_2 and L_2 are state variables at time $t + \delta t$, and $X_2 = X_1 + y_1\delta t$. We need to solve this equation to obtain local loss transition probabilities, $P(t + \delta t, L_2|t, L_1, X_1, y_1)$, which determine the local dynamics of loss variable, L, consistent with conditional loss distributions, $P(t, L|X)$. Similar calculation is performed in [15], where loss transition rates are calculated for unconditional loss distribution. We look for a kernel in the following form

$$P(t + \delta t, L_2|t, L_1, X_1, y_1) = \delta(L_1 - L_2) + \delta t \Lambda(L_2|t, L_1, X_1, y_1), \qquad (4.10)$$

where $\Lambda(L_2|t, L_1, X_1, y_1)$ is a transition intensity from L_1 to L_2. It should satisfy

$$\Lambda(L_2|t, L_1, X_1, y_1) = 0, \quad L_2 < L_1, \qquad (4.11)$$

due to positivity of the loss and

$$\int dL_2 \, \Lambda(L_2|t, L_1, X_1, y_1) = 0, \qquad (4.12)$$

due to probability conservation. Λ satisfies an integral-differential equation,

$$\left(\frac{\partial}{\partial t} + y \frac{\partial}{\partial X}\right) P(t, L|X) = \int dL' \Lambda(L|t, L', X, y) P(t, L'|X). \qquad (4.13)$$

In practice loss distributions, $P(t, L|X)$, are discrete in L. This means that L is a discrete variable, $0 \le L \le L_{max}$. In discrete setting loss densities become vectors of loss probabilities, $P_L(t, X)$. The integral equations above become matrix equations,

$$P_{L_2}(t + \delta t, X_2) = \sum_{L_1=0}^{L_{max}} \left(\delta_{L_2 L_1} + \delta t \, \Lambda_{L_2 L_1}(t, X_1, y_1)\right) P_{L_1}(t, X_1), \qquad (4.14)$$

where $P_L(t, X_2) = P(t, L|X)$ are vectors of loss probabilities, and $\Lambda_{L_2 L_1}(t, X, y)$ are matrices of transition probabilities, which we need to find, satisfying

$$\Lambda_{L_2 L_1}(t, X_1, y_1) = 0, \quad L_2 < L_1,$$

$$\sum_{L_2=0}^{L_{max}} \Lambda_{L_2 L_1}(t, X_1, y_1) = 0. \qquad (4.15)$$

An additional constraint is that loss cannot be bigger then the maximum value, L_{max},

$$\Lambda_{L_{max} L_{max}}(t, X, y) = 0. \qquad (4.16)$$

Therefore, transition probabilities $\Lambda_{L_2 L_1}$ in general have the form

$$\Lambda_{L_2 L_1} = \begin{pmatrix} \Lambda_{00} & 0 & \cdots & 0 & 0 \\ \Lambda_{10} & \Lambda_{11} & \cdots & 0 & 0 \\ \Lambda_{20} & \Lambda_{21} & \cdots & 0 & 0 \\ \vdots & \vdots & \vdots & \vdots & \vdots \\ \Lambda_{L_{max}-1,0} & \Lambda_{L_{max}-1,1} & \cdots & \Lambda_{L_{max}-1,L_{max}-1} & 0 \\ \Lambda_{L_{max},0} & \Lambda_{L_{max},1} & \cdots & \Lambda_{L_{max},L_{max}-1} & 0 \end{pmatrix} \qquad (4.17)$$

The solution of (4.14) is not unique. One needs to provide additional constraints to a find a unique solution. Following [15], we look for solution with bi-diagonal matrix $\Lambda_{L_2 L_1}$

$$
\Lambda_{L_2 L_1} = \begin{pmatrix}
-\Lambda_0 & 0 & \cdots & 0 & 0 \\
\Lambda_0 & -\Lambda_1 & \cdots & 0 & 0 \\
0 & \Lambda_1 & \cdots & 0 & 0 \\
\vdots & \vdots & \vdots & \vdots & \vdots \\
0 & 0 & \cdots & -\Lambda_{L_{\max}-1} & 0 \\
0 & 0 & \cdots & \Lambda_{L_{\max}-1} & 0
\end{pmatrix}
\tag{4.18}
$$

This ansatz means that we are only looking for solutions with localized loss transition probabilities. This is consistent with the picture of just one default per time tick.[e] Constrained in this way transition probabilities are unique and are provided by solution

$$
\Lambda_L(t, y, X) = -\frac{1}{P_L(t, X)} \sum_{L'=0}^{L} \frac{1}{\delta t} \left(P_{L'}(t + \delta t, X + y\delta t) - P_{L'}(t, X) \right).
\tag{4.19}
$$

Alternatively, the same solution can be written as

$$
\Lambda_L(t, y, X) = \frac{1}{P_L(t, X)} \sum_{L'=L_1+1}^{L_{\max}} \frac{1}{\delta t} \left(P_{L'}(t + \delta t, X + y\delta t) - P_{L'}(t, X) \right),
\tag{4.20}
$$

This completes specification of the approximating model for options-like contracts.

4.2. *Pricing of derivatives*

Given the dynamics of the model a derivative contract, $V(t, y, X, L)$, can be priced by backward induction equation

$$
V(t - \delta t, y, X, L) = F\left[\frac{B_{t-\delta t}}{B_t} (1 + \delta t \mathcal{L}) V(t, y, X, L) + f(t, y, X, L) \right],
\tag{4.21}
$$

where $B_{t-\delta t}/B_t$ is the usual discounting factor, operator $\mathcal{L}$ describes dynamics of the model, $f(y, X, L)$ is the source term which describes the contract, and function F describes any non-linear early exercise conditions of the contract. Operator $\mathcal{L}$ describing model dynamics is given by

$$
\mathcal{L}V(t, y, X, L) = \left(\sigma^2 \frac{\partial^2}{\partial y^2} + \mu \frac{\partial}{\partial y} + y \frac{\partial}{\partial X} \right) V(t, y, X, L)
$$

$$
+ \Lambda \int dP(j) \left(V(t, y + j, X, L) - V(t, y, X, L) \right)
$$

$$
+ \Lambda_L(t, y, X) \left(V(t, y, X, L + 1) - V(t, y, X, L) \right),
\tag{4.22}
$$

where Λ_L is loss transition probability, Λ is intensity of (default intensity) jumps and $P(j)$ is distribution of (default intensity) jumps. The first term describes diffusion

---

[e]Such processes are often called *pure-birth processes*, as described in [7].

and drifts, second term describes intensity jumps and the third term describes loss transitions.

A contract is specified by its value at expiry, $V(T, y, X, L)$, the source term which describes the the cash flows of the contract, $f(t, y, X, L)$, and early exercise function $F[V(t, y, X, L)]$. Below we give several examples of contract specifications in this model.

4.2.1. *Vanilla tranches*

To price vanilla tranche one needs to price separately default leg and coupon leg. Default leg is given by pricing a contract in our model with the following specification

$$V(T, y, X, L) = 0, \quad F[V] = V,$$

$$f(t, y, X, L) = \delta t \ \Lambda_L(t, y, X, L) \ (f_T(L + \Delta L) - f_T(L)), \tag{4.23}$$

where T is the maturity of the trade and $f_T(L)$ is the tranche loss function

$$f_T(L) = \max(\min(L, L_E) - L_A, 0), \tag{4.24}$$

where L_A and L_E are attachment and exhaustion of the tranche respectively.

Coupon leg is given by the following choices

$$V(T, y, X, L) = 0, \quad F[V] = V,$$

$$f(t, y, X, L) = c \ \delta t \ (L_E - L_A - f_T(L)), \tag{4.25}$$

where c is the coupon and $f_T(L)$ is the tranche loss function, the same as above. This does not include amortization of the tranche with recovery, which has to be included for super senior tranches.

4.2.2. *European option on tranche*

Option on tranche is a contract where one has a right to buy underlying tranche, V_T, at exercise date T at coupon level, c. Option on tranche is given by the following specification

$$V(T, y, X, L) = \max(V_T(T, y, X, L), 0), \quad F[V] = V,$$

$$f(t, y, X, L) = 0. \tag{4.26}$$

4.2.3. *Leveraged tranche*

In leveraged tranche contract the buyer receives coupon, c, on notional of the contract, N, until one of the triggers is hit. After that the buyer has an option to choose between losing his collateral, L_{cap}, and holding the underlying tranche, V_T, which corresponds to de-leveraging. The contract is specified by

$$V(T, y, X, L) = 0, \quad f(t, y, X, L) = c \, \delta t \, N,$$

$$F[V] = (1 - I(t, y, X, L)) \, V + I(t, y, X, L) \, \max(V_T, -L_{\text{cap}}), \tag{4.27}$$

where $I(t, y, X, L)$ is trigger indicator

$$I(t, y, X, L) = \begin{cases} 1, & \text{if trigger is hit} \\ 0, & \text{if trigger is not hit} \end{cases} \tag{4.28}$$

We see that early exercise function is not just a function of the value of the derivative, V, but it is also a function of value of underlying tranche, V_T. Trigger indicator can depend on t, y, X, L as well as on other underlying contracts.

Let us now discuss the triggers on leveraged tranche. Usually one distinguishes between loss triggers, index spread triggers and mark-to-market of the underlying tranche triggers. It is clear how to define trigger indicator function in case of loss trigger. In case of mark-to-market triggers indicator function simply becomes a function of V_T. In case of index spread triggers, trigger indicator function depends additionally on the index spread of the underlying credit basket corresponding to the current state, $c_{\text{index}}(t, y, X, L)$,

$$c_{\text{index}}(t, y, X, L) = \frac{V_{\text{index default leg}}(t, y, X, L)}{V_{\text{index coupon leg}}(t, y, X, L)}. \tag{4.29}$$

$V_{\text{index default leg}}$ and $V_{\text{index coupon leg}}$ are default and coupon leg (with 100% coupon) of the underlying index, which can be calculated on the same lattice as described in Sec. 4.2.1.

To summarize, to price leveraged tranche one needs to first calculate value of the underlying tranche, V_t, value of the index default and coupon legs, $V_{\text{index default leg}}$ and $V_{\text{index coupon leg}}$, for every value of t, y, X, L. Then one can calculate leverage tranche value, V, on the same lattice, using specification (4.27), with trigger indicator function in general dependent on L, V_T, $V_{\text{index default leg}}$ and $V_{\text{index coupon leg}}$.

4.2.4. Tranche with counterparty risk

Tranche with counterparty default risk, V, is a derivative with vanilla tranche, V_T as underlying. It is given by the following specification

$$V(T, y, X, L) = V_T(T, y, X, L), \quad f(t, y, X, L) = f_T(t, y, X, L),$$

$$F[V] = (1 - \delta t \lambda_c) V + \delta t \lambda_c F_{\text{pay-on-default}}[V_T], \tag{4.30}$$

where λ_c is intensity of the counterparty default corresponding to the current state, $\lambda_c = \lambda_c(t, y, X, L)$, and $F_{\text{pay-on-default}}$ is the payment on default of the counterparty, which depends on the current mark-to-market of the underlying tranche, V_T. Two possible choices for $F_{\text{pay-on-default}}$ are

$$F_{\text{pay-on-default}} = 0, \tag{4.31}$$

which correspond to no payments, and

$$F_{\text{pay-on-default}} = \min(V_T, L_{\text{cap}}), \tag{4.32}$$

which corresponds to counterparty paying mark-to-market of the underlying tranche only up to a certain cap L_{cap}. What remains is to determine the intensity of the counterparty default, λ_c, corresponding to the current state. As discussed in Sec. 2 survival probability conditioned on realized intensities depends only on variable X, $p_c(t, y, X, L) = p_c(t|X)$, and is given by

$$p_c(t|X) = e^{-\int_0^t \lambda_i dt} = p_c(t)e^{-\beta_c(t)X + \phi(t, \beta_c(t))}, \tag{4.33}$$

$$\beta_c(t)t = \int_0^t \lambda_c^f dt, \tag{4.34}$$

where $p_c(t)$ is implied survival probability, and λ_c^f is the forward intensity corresponding to $p_c(t)$. Similarly to the calculate of the loss transition probabilities in Sec. 4.1, we can calculate the intensity of the counterparty default, λ_c

$$\lambda_c(t, y, X, L) = -\frac{1}{p_c(t|X)} \frac{1}{\delta t} \left(p_c(t + \delta t | X + y\delta t) - p_c(t|X) \right). \tag{4.35}$$

This completes the definition of the problem and tranche with counterparty risk can be priced by backward induction.

5. Conclusions

In this paper we discussed a stochastic intensity model in the context of pricing exotic structured credit derivatives. The model is defined in terms of the microscopic local dynamics of intensities of individual names and therefore can be used for pricing a wide range of derivatives. We discussed various parameterisations of the model with a view to find parameterisations, which are as economical as possible, while still containing relevant degrees of freedom to be useful for pricing. From the point of view of vanilla tranche pricing the model induces one-factor copula for individual credit defaults. This provides a bridge between stochastic intensity modelling and more conventional default modelling with copulas. This also allows a variety of techniques developed for copula models to be used for the stochastic intensity model discussed here.

Efficient calculation of vanilla tranches allows one to attempt a brute force calibration of the model dynamics to the standard tranche CDO market. We find that the model, even in its simplest specification, contains relevant degrees of freedom and is flexible enough to calibrate well to separate maturities. More sophisticated parametrizations of the model dynamics allows one to calibrate reasonably well to the standard tranches of different maturities and in this way effectively model the term structure of the correlation skew. High-precision calibration to different instruments with different maturities will benefit from more advanced techniques.

We also discussed the pricing of various exotic derivative contracts in the framework of stochastic intensity models. As the model is specified microscopically for individual credits many contracts can be priced by Monte-Carlo simulation. Option-like contracts, which require backward induction for pricing, can be priced in a low-dimensional effective model for loss with dynamics induced by the full stochastic intensity model.

Acknowledgments

We would like to thank our colleagues at Merrill Lynch for fruitful discussions and useful insights.

References

[1] L. Andersen and J. Sidenius, Extensions to the gaussian copula: Random recovery and random factor loadings, working paper (2004).

[2] M. Baxter, Levy process dynamic modelling of single-name credits and CDO tranches, working paper (2006).

[3] T. Björk, Y. Kabanov and W. Runggaldier, Bond market structure in the presence of marked point processes, working paper (1996).

[4] D. Brigo, A. Pallavicini and R. Torresetti, Calibration of CDO tranches with the dynamical generalized-poisson loss model, working paper.

[5] X. Burtschell, J. Gregory and J.-P. Laurent, Beyond the gaussian copula: Stochastic and local correlation, working paper (2005).

[6] D. Duffie and N. Garleanu, Risk and valuation of collaterized debt obligations, *Financial Analysts Journal* **57**(1) (2001) 41.

[7] W. Feller, *An Introduction to Probability Theory and its Applications*, Vol. 1 (1968).

[8] R. Gaspar and T. Schmidt, Term structure models with shot noise effects, working paper.

[9] P. den Iseger, Numerical transform inversion using gaussian quadrature, working paper (2006).

[10] M. Joshi and A. Stacey, Intensity gamma: A new approach to pricing credit derivatives, working paper (2005).

[11] D. Li, On default correlation: A copula function approach, *Journal of Fixed Income* **9** (2000) 43.

[12] F. A. Longstaff and E.S. Schwartz, Valuing american options by simulation: a simple least-squares approach, working paper (2001).

[13] L. McGinty and R. Ahluwalia, Introducing base correlation, A model for base correlation calculations, working paper (2004).

[14] A. Mortensen, Semi-analytical valuation of basket credit derivatives in intensity-based models, working paper (2005).

[15] P. J. Schnbucher, Portfolio Losses and the Term Structure of Loss Transition Rates: A new methodology for the pricing of portfolio credit derivatives, working paper (2005).

[16] P. J. Schnbucher and D. Schubert, Copula-dependent default risk in intensity models, working paper (2001).

[17] J. Sidenius, V. Piterbarg and L. Andersen, A new framework for dynamic credit portfolio loss modelling, working paper (2005).

[18] P. A. C. Tavares, T. Nguyen, A. Chapovsky and I. Vaysburd, Composite basket model, working paper (2004).

LARGE PORTFOLIO CREDIT RISK MODELING

MARK H. A. DAVIS

Department of Mathematics, Imperial College London
London SW7 2AZ, United Kingdom
mark.davis@imperial.ac.uk

JUAN CARLOS ESPARRAGOZA-RODRIGUEZ

Milliman Inc. 103 Bunhill Row
London EC1Y 8LZ, United Kingdom
juan.esparragoza@milliman.com

A model for large portfolio credit risk is developed by using results on the asymptotic behavior of stochastic networks. An efficient pricing technique is proposed using a newly-introduced quadrature algorithm. Accurate calibration to iTraxx tranche spreads is demonstrated.

Keywords: Stochastic network; functional law of large numbers; functional central limit theorem; quadratures.

1. Introduction

Rating based models have been widely used for pricing and risk management purposes. Maybe the best known is the Jarrow, Lando and Turnbull model [12]. More recent application of rating models in the context of Markov chains can be found in, for example, Frydman and Schuermann [7] and Christensen *et al.* [3].

Even if there are some criticisms of the use of credit ratings for pricing purposes, it is appealing to consider having buckets of obligors with similar characteristics to simplify the mathematics while modeling the likelihood of obligors to modify their credit worthiness. This is particularly useful when handling large portfolios.

The use of latent variables and common factors has been common practice when dealing with large portfolios. Examples well known are the CreditMetrics model [10] and PortfolioRisk+ methodology [4]. Moreover, empirical evidence [8] suggests that the pattern of realized defaults is well represented by a latent variable model that postulates the existence of an exogenous process interpreted as a 2-state (growth/recession) economic variable, although introduction of further states can increase the explanatory power.

Finally, taking an approach of ratings and latent variables related with economic interpretation, it is sensible to assume that obligors move around rating categories at a faster time scale than the economic cycle.

These facts suggest a model in which obligors move around the rating categories at rates depending on the latent process and occasionally default. Moreover, there is an obvious analogy with stochastic networks in which "clients" move around "service stations" for processing. There is a huge literature on the analysis of stochastic networks — see Whitt [21] for a recent textbook account. In particular, work by Choudhury *et al.* [1] studies fluid and diffusion limits for stochastic networks under random environments. We give some modifications of their results in this paper, in a form suitable for application to credit portfolios.

The paper is laid out as follows. Section 2 describes our model. Section 3 gives a summary of the mathematical results on fluid and diffusion limits that we need. The application of these results to our model is given in Sec. 4. We then turn to computational matters. Section 5 describes our main computational algorithm based on quadrature formulas and compares it with conventional Monte Carlo approaches. In Sec. 6 we give the results of calibration studies where the data consists of iTraxx index tranche spreads on several dates. We demonstrate that accurate calibration is possible if our latent variable is chosen properly. Concluding remarks are given in Sec. 7.

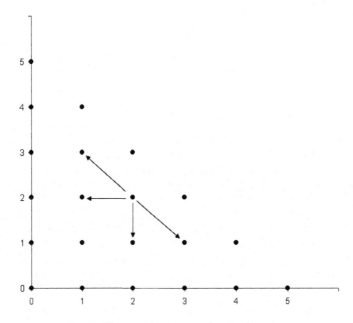

Fig. 1. The sample space in the simplest case.

2. Model Description

Assume a portfolio with n obligors and m possible ratings $1, \ldots, m$. The initial rating composition of the portfolio is represented by the *rating distribution vector* $q^n(0) \in \mathbb{R}^m$. The vector $q^n(t)$ will represent the random rating distribution of the portfolio at a later time t. This is, $(q^n(t))_i$ is the number of obligors in the rating category i at time t (rather than a proportion). We assume the transition probabilities are the same for all obligors in the same rating class. Then, $\sum q_k^n(t)$ is the number of non defaulted obligors at time t. When $n = 5, m = 2$, the state space of $q^n(t)$ is as shown in Fig. 1. The figure shows all the possible movements given the current credit ratings: transitions (moves along the diagonal) and defaults (moves to the next diagonal).

We assume as well there is a finite-state random environment process $\{\xi_t, t \geq 0\}$ representing some macroeconomic (or sector associated) process that influences the default/transition rates of the obligors. The obligor credit events are independent conditional on the realization of the environment process and follow a Markov chain with default/transition rates being defined as a function of the environment process. The dependence structure of the defaults in the portfolio comes from this environment process.

The finite state environment process defines different "layers" in which transition parameters are different. These multiple layers are copies of the initial state space shown in Fig. 1. Now the process jumps across the layers whenever the random environment process ξ_t jumps to a different state as illustrated in Fig. 2. In this model the movements across the grid are expected to be more frequent than the movements across layers, particularly when the number of elements in the system

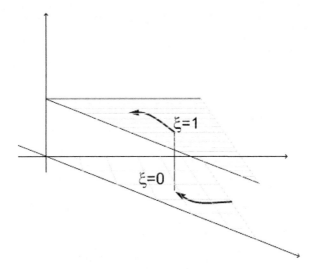

Fig. 2. Multiple layer setting.

increases. This is why in some literature this structure is referred to as a slowly changing environment [15]. All we can observe however is the second coordinate of the pair process $(\xi_t, q^n(t))$. For the purposes of this paper we will consider the environment process ξ_t being a Markov chain with an associated matrix Q, however some of the results hold for arbitrary processes.

Each credit events occurs as the first jump of a Poisson process with rate dependent of the state of the environment process ξ_t. Transition rates from i to j are denoted $\mu_{ij}(\xi)$, $i, j = 1, \ldots, m$, $i \neq j$. Default rates are denoted $\mu_i(\xi) = \mu_{ii}(\xi)$, $i = 1, \ldots, m$. Then, a credit event occurs according to the first jump of a Poisson process with rate

$$\hat{\mu}_i(\xi) = \sum_{j=1}^{m} \mu_{ij}(\xi) \tag{2.1}$$

Once a credit event occurs at time t, the obligor defaults with probability $\mu_i(\xi)/\hat{\mu}_i(\xi)$, while a transition to rate $j \neq i$ has probability $\mu_{ij}(\xi)/\hat{\mu}_i(\xi)$.

From the description above, for $n > 0$ the process $q^n(t)$ can be expressed as the solution to the following integral equation

$$q^n(t) = q^n(0) + \sum_{i=1}^{m} \sum_{j=1}^{m} N_{ij} \left(\int_0^t \mu_{ij}(\xi_s) q_i^n(s) ds \right) v_{ij} \tag{2.2}$$

where for $i, j = 1, \ldots, m$, N_{ij} is a collection of independent Poisson processes, and v_{ij} is the set of vectors defined as

$$v_{ij} = \begin{cases} \mathbf{e}_j - \mathbf{e}_i & j \neq i \\ -\mathbf{e}_i & i = j \end{cases} \tag{2.3}$$

The generator matrix $A(\xi)$ associated to the process is defined as

$$A_{ij}(\xi) = \begin{cases} \mu_{ij}(\xi) & j \neq i \\ -\hat{\mu}_i(\xi) & i = j \end{cases} \tag{2.4}$$

2.1. Formal definition of the model

The random environment process ξ_t is a finite state process in continuous time taking values in a set of indices Ξ and having at most a finite number of jumps in any bounded interval of $[0, \infty)$.

To construct the process $q^n(t)$ we consider a collection of mutually independent Poisson processes $\{N_i\}_{i \in I}$, for a set of indices denoted by I. We define a collection of vectors $\{\mathbf{v}_i\}$ in $\mathbb{R}^m$, $m \in \mathbb{N}$, and a collection of non-negative functions of the form $\mu_{i,n}(\cdot, \cdot, \xi) : [0, \infty) \times \mathbb{R}^m \to [0, \infty)$ for all $i \in I$ and $\xi \in \Xi$. We assume each $\mu_{i,n}(t, \cdot, \xi)$ is continuous with respect to the second argument. More generally we can assume the function being Lipschitz bounded (see [5]), however for the purpose of this paper continuity is always assumed.

We define the probability space $(\Omega^1, \mathfrak{F}^1, P^1)$ associated to the random environment process ξ_t, where $\Omega^1 = D([0, \infty), \Xi)$ the space of Right Continuous with Left Limits (RCLL) functions from $[0, \infty)$ into Ξ, $\mathcal{F}^1$ is a σ-algebra in Ω^1 and P^1 is some probability measure. The measure space associated to the collection $\{A_i\}$ is defined by the state space $\Omega^2 = \Omega_1 \times \cdots \times \Omega_p$ where $\Omega_i = D([0, \infty), \mathbb{R}^n)$ and the minimum sigma-algebra $\mathfrak{F}^2$ generated by Ω^2. The probability measure P^2 is defined as the product measure. Finally, we assume independence between the environment process and each Poisson process A_i, therefore the probability space is defined as $(\Omega, \mathfrak{F}, P)$ with $\Omega = \Omega^1 \times \Omega^2$, $\mathfrak{F} = \sigma(\Omega)$ and $P = P^1 \times P^2$. We will denote by $\omega = (\omega_1, \omega_2)$ the elements of Ω being $\omega_1 \in \Omega^1$ and $\omega_2 \in \Omega^2$.

We define a mapping Υ from Ω into $D([0, \infty), \mathbb{R}^m)$ by $(\omega_1, \omega_2) \mapsto q$, where q^n is the process solution to the equation

$$q^n(t) = q^n(0) + \sum_{i \in I} N_i \left(\int_0^t \mu_{i,n}(s, q^n(s), \xi_s) ds \right) \mathbf{v}_i \qquad (2.5)$$

The law of $q^n(t)$ is $P_n(\cdot) = P(\Upsilon^{-1}(\cdot))$. The probability conditional on the environment process ξ_t denoted by $P_n^{\omega_1}$ is then defined by the conditional probability under the inverse mapping. Such processes is well defined (has a unique solution) as shown in Theorem 2.3.1 in [5].

In the following we will assume the random environment process ξ_t is a finite-state Markov chain. Then we can characterize P^1 by the generator matrix Q that defines the process.

3. Fluid and Diffusion Limits

The model as enunciated above can become numerically cumbersome as the dimension of the portfolio increases. However, relatively accurate and tractable analysis of the model can be done by studying the asymptotic behavior of $q^n(t)$ as $n \to \infty$ conditional on the realization of the environment process ξ_t and using an appropriate scaling.

The model can be considered as a stochastic network where obligors represent clients in a service network that transit across different service stations (ratings) with infinite servers. Once a client is served it transits to a different service station (credit migration) or leaves the system (default). In this context $q^n(t)$ represent the state of the stochastic network at time t.

There is a wide interest in the convergence properties of queues and stochastic networks (queueing systems) according to different scalings — see Whitt [21]. A Functional Strong Law of Large Numbers (FSLLN) is derived for sequences of the form $q^n(t)/n$, this is commonly referred as the *fluid limit*. This strong convergence result is usually expressed as the unique solution to a partial differential equation and it helps to study the stability of the system. With the fluid limit $q^{(0)}(t)$ a weak convergence result can also be derived for sequences of the form

$$\frac{1}{\sqrt{n}}\left(q^n(t) - q^{(0)}(t)\right)$$

and is known as Functional Central Limit Theorem (FCLT) or *diffusion limit*. The FCLT gives a diffusion process that refines the approximation (in distribution) that may be obtained by the fluid limit.

Studies of the fluid and diffusion limits for queue applications can be traced to early works such as Iglehart and Whitt [11] and is still an area of development. An important branch of the development is based on heavy traffic condition assumptions, meaning an equilibrium between arrivals and service rates, leading to a stationary equilibrium in the state of the system. A less restrictive approach can be used when the system is Markov whereby the use of random times convergence results are obtained, these kind of results has been studied in Kurtz [13] and Ethier and Kurtz [6], more recently Mandelbaum *et al.* [15] present results applicable to Markovian stochastic networks. Specific applications to the credit model explained above are presented below.

Slowly changing environments are processes within a system that change at a slower rate than the stochastic network or queue influencing its performance, that is the case of the random environment process in our credit model. An example is when the rate of service depends on an independent process and is kept constant when the number of clients increases. Choudhury *et al.* [1] show that for an infinite capacity server under slowly changing environments is possible to derive fluid and diffusion limits conditional on the realization of the environment process. In this case, the fluid limit is the solution to a PDE whose parameters depend on the environment process. The diffusion limit is a Brownian diffusion conditional on the environment process.

In [5] we show that some of the results presented by [15] hold under slowly changing environments without any heavy traffic assumption. In particular the diffusion limit is given by a diffusion with stepwise constant diffusion coefficient and therefore the process can be completely defined by its covariance matrix obtained as the solution to the Lyapunov equation.

4. Convergence Results for the Rating Distribution Process

In the following we present the convergence results for the credit model described in Sec. 2. These are applications of results presented in [5] and detailed proofs can be found there. These results are modifications of results presented by Choudhury *et al.* [1] concerning queueing networks.

4.1. *The fluid limit*

Theorem 4.1. *Assume we have a sequence of rating processes (ξ, q^n) $n = 1, 2 \ldots$ according to the Markov environment process ξ_t and to the rating process* (2.2).

Suppose that

$$\frac{q^n(0)}{n} \to q^{(0)}(0) \quad \text{as } n \to \infty. \tag{4.1}$$

Then for each $\omega_1 \in \Omega^1$

$$\frac{q^n(t)}{n} \to q^{(0)}(t) \quad \text{as } n \to \infty \tag{4.2}$$

a.s. in $P_Q^{\omega_1}$, *where* $q^{(0)}(t)$ *defined in* $\mathbb{R}^m$ *is the solution to the PDE system*

$$\frac{d}{dt} q^{(0)}(t) = A(\xi_t) q^{(0)}(t) \tag{4.3}$$

that is deterministic conditional on the realization of ξ. *Here the matrix-valued function A is given by* (2.4).

This theorem is a special case of Theorem 2.3.2 in [5]. Since ξ has at most finitely many jumps in any bounded interval of time, we can define the countable set of jump times of ξ, $t_0 = 0, t_1 < \cdots$, and define $q^n(t)$ recursively as

$$q^n(t) = e^{(t-t_i)A(\xi_{t_i})} q^n(t_i) \tag{4.4}$$

for $t_i < t \leq t_{i+1}$.

The function $q^{(0)}(t)$ is called the *fluid limit* of the sequence $\{q^n(t)\}_{n \geq 0}$.

4.2. *The diffusion limit*

Theorem 4.2. *Assume we have a sequence of rating processes* (ξ, q^n) $n = 1, 2 \ldots$ *according to the Markov environment process* ξ_t *and to the rating process 2.2. For* P^1 *a.e.* $\omega_1 \in \Omega^1$, *if*

$$\lim_{n \to \infty} \sqrt{n} \left[\frac{q^n(0)}{n} - q^{(0)}(0) \right] =^d q^{(1)}(0) \tag{4.5}$$

w.r.t $P_Q^{\omega_1}$, *then*

$$\lim_{n \to \infty} \sqrt{n} \left[\frac{q^n(t)}{n} - q^{(0)}(t) \right] =^d q^{(1)}(t) \tag{4.6}$$

w.r.t $P_Q^{\omega_1}$, *where the process* $q^{(1)}(t)$ *takes values in* Ω^2 *and it is the solution to the stochastic integral equation*

$$q^{(1)}(t) = q^{(1)}(0) + \int_0^t A_t q^{(1)}(t) + \sum_{k=1}^m \int_0^t \left(q_k^{(0)}(t) \right)^{1/2} B_k dW_t^{(l)} \tag{4.7}$$

where W^l *is a m-dimensional vector of independent standard Brownian motions for* $k = 1, \ldots, m$, *and* $A_t = A(\xi_t)$. *The matrices* B_k *have components*

$$(B_k(t))_{ij} = \begin{cases} -\mu_{ij}^{1/2} & \text{if } i = j \\ \mu_{ij}^{1/2} & \text{if } k = i \neq j \\ 0 & \text{otherwise} \end{cases} \tag{4.8}$$

This is, conditional on the realization of ξ the limit rating process $q^{(1)}(t)$ is normally distributed for any $t > 0$.

The theorem can be seen as a special case of Theorem 2.3.3 in [5].
It is always possible to rewrite (4.7) as

$$q^{(1)}(t) = q^{(1)}(0) + \int_0^t A_t q^{(1)}(t) + \int_0^t B(s) d\hat{W}_t^{(k)}$$

where $\hat{W}_t$ is a m-dimensional vector of independent standard Brownian motions. In the case $m = 2$ the SDE is

$$dq_1^{(1)}(t) = -q_1^{(1)}(t)(\mu_1(t) + \mu_{12}(t))dt + q_2^{(1)}(t)\mu_{12}(t)dt - (q_1^{(0)}(t))^{1/2} \quad (4.9)$$

$$\times(\mu_1^{1/2}(t)dW_{1,t}^{(1)} + \mu_{12}^{1/2}(t)dW_{2,t}^{(1)}) + (q_2^{(0)}(t)\mu_{21}(t))^{1/2}dW_{1,t}^{(2)}$$

$$dq_2^{(1)}(t) = -q_2^{(1)}(t)(\mu_2(t) + \mu_{21}(t))dt + q_1^{(1)}(t)\mu_{21}(t)dt - (q_2^{(0)}(t))^{1/2}$$

$$\times(\mu_2^{1/2}(t)dW_{2,t}^{(2)} + \mu_{21}^{1/2}(t)dW_{1,t}^{(2)}) + (q_1^{(0)}(t)\mu_{12}(t))^{1/2}dW_{2,t}^{(1)}$$

That is equivalent to the the following SDE system

$$dq^{(1)}(t) = A_t q^{(1)} dt + B(t) d\hat{\mathbf{W}}_t \quad (4.10)$$

where

$$B(t) = \begin{pmatrix} \sigma_1(t) & 0 \\ \rho(t)\sigma_2(t) & \sqrt{1 - \rho^2(t)}\sigma_2(t) \end{pmatrix} \quad (4.11)$$

$$\sigma_1^2(t) = q_1^{(0)}(t)(\mu_1(t) + \mu_{12}(t)) + q_2^{(0)}(t)\mu_{21}(t) \quad (4.12)$$

$$\sigma_2^2(t) = q_2^{(0)}(t)(\mu_2(t) + \mu_{21}(t)) + q_1^{(0)}(t)\mu_{12}(t) \quad (4.13)$$

$$\rho(t)\sigma_1(t)\sigma_2(t) = -q_1^{(0)}(t)\mu_{12}(t) - q_2^{(0)}(t)\mu_{21}(t) \quad (4.14)$$

and

$$\hat{\mathbf{W}}_t = (\hat{W}_t^{(1)}, \hat{W}_t^{(2)})^t \quad (4.15)$$

is a standard Brownian motion in $\mathbb{R}^2$.

In general we can derive that the matrix $\hat{B} = B'B$ has components

$$\hat{B}_{ij}(t) = \begin{cases} q_i^{(0)}(t)\sum_k \mu_{ik} + \sum_k q_k^{(0)}\mu_{ki} & \text{if } i = j \\ -q_i^{(0)}\mu_{ij} - q_j^{(0)}\mu_{ji} & \text{otherwise} \end{cases} \quad (4.16)$$

On the interval $[t_i, t_{i+1})$, $A_t = A$ is a constant matrix equal to $A(\xi_{t_i})$ and hence the solution of the SDE is given by

$$q^{(1)}(t) = e^{(t-t_i)A}q^{(1)}(t_i) + \int_{t_i}^{t} e^{(t-s)A}B(s)d\hat{\mathbf{W}}_s \tag{4.17}$$

with $q^{(1)}(0) = \mathbf{0}$. The process is a stable Gaussian system with covariance matrix given by the integral

$$C_t = \mathrm{Cov}[q^{(1)}(t), q^{(1)}(t)] = e^{(t-t_i)A}C_{t_i}e^{(t-t_i)A'} + \int_{t_i}^{t} e^{(t-s)A}B(s)B(s)'e^{(t-s)A'}ds \tag{4.18}$$

that can be calculated numerically by solving the Lyapunov matrix ordinary differential equation

$$\frac{d}{dt}C_t = AC_t + C_tA' + B(s)B'(s), \tag{4.19}$$

with the known initial condition at $t = t_i$. This has a unique solution equal to the expression in (4.18).

Similarly to the fluid limit, conditional on the realization of ξ_t and applying (4.17)–(4.19) piecewise we can express the solution as

$$q^{(1)}(t) = e^{(t-t_i)A_{t_i}}q^{(1)}(t_i) + \int_{t_i}^{t} e^{(t-s)A_{t_i}}B(s)d\hat{W}_s \tag{4.20}$$

for $t_i < t < t_{i+1}$ where t_i is the time of the ith jump of ξ_t.

Conditional in the realization of the random environment, the process $q^n(t)$ may be approximated by the two processes

$$q^n(t) \simeq nq^{(0)}(t) + \sqrt{n}q^{(1)}(t)$$

where $q^{(0)}(t)$ is deterministic and $q^{(1)}(t)$ is a diffusion called the "diffusion limit" of the sequence $q^n(t)$. This is, conditional on the random environment, the distribution of the process $q^n(t)$, at any time $t > 0$, may be approximated by a normal distribution.

4.3. *The infinitesimal generator of the single-obligor process and the probability of default*

In this section we derive a the probability of default for a single obligor at time $t > 0$. This is important for calibration purposes as it allows us to price CDS in a closed form.

To obtain the probability of default of a single bond consider q_t^1 the rating distribution process of a single bond portfolio under the random environment process ξ_t and initial rating $1 \leq k \leq m$. The pair process (ξ_t, q_t^1) is then a Markov chain in continuous time with space state $\Xi \times (\{\mathbf{e_i}; i = 1, \ldots, m\} \cup \mathbf{0}\})$ and transition rates defined as

$$\lambda((\eta, \mathbf{e}_i) \to (\eta, \mathbf{e}_j)) = \mu_{ij}^{\eta} \qquad \text{for credit rate transitions,}$$
$$\lambda((\eta, \mathbf{e}_i) \to (\zeta, \mathbf{e}_i)) = Q_{\eta\zeta} \qquad \text{for environment state transitions,}$$
$$\lambda((\eta, \mathbf{e}_i) \to (\eta, \mathbf{0})) = \mu_i^{\eta} \qquad \text{for transition to default,}$$
$$\lambda((\xi, \mathbf{0}) \to (\zeta, \mathbf{0})) = Q_{\eta\zeta} \qquad \text{for environment state transitions once}$$
$$\text{default has occurred.}$$

$\eta, \zeta \in \Xi$ and $i, j = 1, \ldots, m$.

The default states $(\xi, \mathbf{0})$ are absorbent. Alternatively, we can think of the process as a Markov chain in the space $\Xi \times \{\mathbf{e}_i; i = 1, \ldots, m\}$ killed at a rate μ_i^{ξ} in state $(\xi, \mathbf{e}_i)$ and defaults corresponding to the cemetery state of the process.

The default probability at time t is then given by

$$\mathbb{E}\{\mathbb{I}_{\tau \geq t}\} = \int_0^t P_s\Big(1 - \sum q_i^1(t)\Big) ds$$

where P is the infinitesimal generator of the pair process (ξ, q_t^1).

Considering the process $q^1(t)$ under the environment state $\xi_t = \xi$ we know that the generator matrix $A_\xi \in \mathbb{R}^{m+1 \times m+1}$ is given by

$$(A_\xi)_{ij} = \begin{cases} -\sum_{k \neq i} \mu_{ik}^{\xi} - \mu_i^{\xi} & i = j \leq m \\ \mu_{ji}^{\xi} & i \neq j, i, j \leq m \\ \mu_j^{\xi} & i = m+1, j \leq m+1 \\ 0 & \text{otherwise} \end{cases} \qquad (4.21)$$

and from there it is possible to derive the generator of the process (ξ_t, q_t^1).

Proposition 4.1. *The infinitesimal generator of the single obligor process (ξ, q_t^1) is the linear operator $A \in \mathbb{R}^{n_\xi(m+1) \times n_\xi(m+1)}$ defined with the matrices A_ξ, $\xi \in \Xi$, as*

$$A = \begin{pmatrix} A_0 - \lambda_0 I & \lambda_1 Q_{10} I & \cdots & \lambda_{n_\xi-1} Q_{n_\xi-1,0} I \\ \lambda_1 Q_{01} I & A_1 - \lambda_1 I & \cdots & \lambda_{n_\xi-1} Q_{n_\xi-1,0} I \\ \vdots & \vdots & \ddots & \vdots \\ \lambda_0 Q_{0n_\xi} I & \lambda_0 Q_{1n_\xi} I & \cdots & A_{n_\xi} - \lambda_{n_\xi} I \end{pmatrix}.$$

Here n_ξ is the cardinality of Ξ and $\lambda_i = \sum_j Q_{ij}$.

With the infinitesimal generator the probability transition matrix can be calculated by simple integration and the probabilities of default can be calculated.

Proposition 4.2. *The probability of default in a horizon time t for an obligor with initial rate i is given by*

$$\mathbb{E}\{\mathbb{I}_{\tau \geq t}\} = \sum_{k=1}^{n_\xi} \left(e^{tA}\right)_{k(m+1),i} \tag{4.22}$$

where A is the infinitesimal generator of the single obligor process.

Proof. The probability of the time to default τ being less t is the probability that the process is in a state $(\xi, \mathbf{0})$ at time t. Then the probability is given by the sum of the entries of the transition matrix e^{tA}. $\square$

5. Computational Aspects: Quadratures

Some of the computational tools available for pricing general claims, where analytical formulas are not straightforward, are the use of a numerical solutions to the backward or forward equations and procedures based on sampling the environment process such as Monte Carlo. When solving the Kolmogorov equations numerically, the problem can be reduced to handling a PDP system, however the solution can be computationally expensive.

We will consider a general European claim on q_t^n, the rating/default distribution of the portfolio with n elements exercised at time T. Let us denote the exercise value of this claim as $f(q_T^n)$. In the case of the losses in the portfolio, for example, the form of the claim is $f(x) = n - \sum x_i$. Since we assume that the associated diffusion limit $q_t = nq_t^{(0)} + \sqrt{n}q_t^{(1)}$ approximates q_t^n "reasonable well", we will assume as well that f is "smooth enough" to allow for the approximation of $f(q_t^n)$ by $f(q_t)$.

A natural approach to calculate expectations is by conditioning on the environmental process path $\{\xi_t : 0 \leq t \leq T\}$, then

$$\mathbb{E}\{f(q_T)\} = \mathbb{E}\left[\mathbb{E}\{f(q_T)|\xi_t : 0 \leq t \leq T\}\right]. \tag{5.1}$$

As we are concerned with pricing, in the following we will assume $\mathbb{E}$ is the expectation under a risk neutral measure $\mathbf{Q}$ unless stated differently.

The advantage of this idea is that q_t conditional on the environment path is jointly normal distributed and expectations are straightforward to calculate, depending solely in the expectation (fluid limit) and variance of the process (given by the Lyapunov equation in Sec. 4). The remaining choice is whether to apply Monte Carlo methods to generate paths $\{\xi_t|0 \leq t \leq T\}$ or to approximate the distribution of the jump times t_i and apply numerical integration. While the former method requires little programming effort and gives goods results, a refined version of sampling can give even better results.

We will consider the process $\{\xi_t : 0 \leq t \leq T\}$ and N_T, the associated counting process of the number of jumps in ξ_t in interval $[0, T]$. Assuming the environment process takes only two values, the process ξ_t can be characterized by the initial state ξ_0 and the vector of jump times $\underline{t} = (t_1, \ldots, t_N) \in [0, T]^n$ where $t_N < T$ and

$t_{N+1} > T$, corresponding to $N_T = N$. Alternatively, it can be defined in terms of the vector of inter-jump times $\underline{\tau} = (\tau_1, \ldots, \tau_n) \in S_T^n$ where $\tau_i = t_i - t_{i-1}$ with $t_0 = 0$. We denote by S_T^n the n-dimensional simplex set defined by the points $\underline{\tau}$ such that $\sum \tau_i \leq T$. In the general case we must consider all possible paths as we explain later, but to simplify the exposition we will assume a two-state environment process in the remainder of this section unless otherwise stated.

In the case of European claims of the form $f(q_T)$, the conditional variable $f(q_T)|\{\xi_t; 0 \leq t \leq T\}$ can be expressed in terms of the vector $\underline{t}$ and will be denoted as $f(q_T; \underline{t})$. Alternatively, it can be expressed in terms of the time between jumps, $\tau_i = t_i - t_{i-1}$ for $i = 1, \ldots, N$ and $t_0 = 0$, denoted as $f(q_T; \tau)$. The vectors $\underline{t}$ and τ are random vectors which we will characterize explicitly later.

Using a naive Monte Carlo approach, a first version of our pricing algorithm is:

1. For $i = 1$ to M the number of iterations.

   (a) Simulate a sequence of jumps $t_1, \ldots, t_{N_i}$ and the process ξ_t, such that $t_{N_i} < T$ and $t_{N_i+1} \geq T$, according to the rate $\lambda(t_{j-1})$, $j = 1, \ldots, N_i + 1$ and the transition matrix Q.
   (b) Based on the simulated process ξ_t generate the parameters $\mu(\xi)$ and C_t of the diffusion process q_t.
   (c) Calculate $f(T) = \mathbb{E}\{f(q_T)|\xi_t : 0 \leq t \leq T\}$

2. Average the results to obtain $\mathbb{E}\{f(q_T)\}$

This algorithm has some problems, the most obvious is that if the jump rates of the environment process ξ_t are low, a large proportion of samples will show no jumps at all, so that the number of samples required to obtain convergence will be large.

An alternative is to use Importance Sampling techniques where a measure, say $\tilde{\mathbf{Q}}$, is chosen to increase the frequency of jumps in the environment process. Hopefully this will improve the convergence in the pricing, particularly for senior tranches. Under the Importance Sampling scheme, the basic algorithm is modified as follows:

1. For $i = 1$ to M the number of iterations.

   (a) Simulate a sequence of jumps $t_1, \ldots, t_{N_i}$ and the process ξ_t, such that $t_{N_i} < T$ and $t_{N_i+1} \geq T$, according to the rate $\tilde{\lambda}(t_{j-1})$, $j = 1, \ldots, N_i + 1$ and the transition matrix $\tilde{Q}$ consistent with the measure $\tilde{\mathbf{Q}}$.
   (b) Based on the simulated process ξ_t generate the parameters $\mu(\xi)$ and C_t of the diffusion process q_t and the Radon–Nikodym derivative.
   (c) Calculate

   $$f(T) = \mathbb{E}\{f(q_T)|\xi_t : 0 \leq t \leq T\}$$

   and the Radon–Nikodym derivative $\frac{d\mathbf{Q}}{d\tilde{\mathbf{Q}}}$, see Rogers and Williams [18] for a study of change of measures in Markov Chains.

2. Average the results to obtain

$$\mathbb{E}_{\tilde{\mathbf{Q}}}\left\{f(q_T)\frac{d\mathbf{Q}}{d\tilde{\mathbf{Q}}}\right\} = \mathbb{E}\left\{f(q_T)\right\}$$

We propose an alternative technique that combines this conditional analysis with the use of Gaussian Quadratures.

5.1. *CDO pricing*

In principle we are concerned with the loss process at time T up to the attachment point of the tranche K (we consider base tranches as building blocks [16]). In this section we will consider the limit process $q(t) = q_t^{(0)} + \frac{1}{\sqrt{n}}q_t^{(1)}$, this is, instead of absolute losses we will consider losses as the proportion of the original notional. In the CDO case the claims can be expressed as

$$f(q_T) = \min\left\{\left(1 - \sum_i q_i(T)\right)(1-r), K\right\} = L_t \wedge K$$

where r is the deterministic recovery rate. We will denote the loss process $L_T = (1 - \sum_i q_i(T))(1-r)$.

Assuming a deterministic discount function $D_i = D(t_i)$ and that the payment of the losses occurs at the end of the coupon period, the spread u of the CDO is the solution to the equation

$$u\tau \sum_{i=0}^{N-1} D_i \mathbb{E}\left(K - f(q_{t_i})\right)^+ = \sum_{i=1}^{N} D_i \mathbb{E}(f(q_{t_i}) - f(q_{t_{i-1}})) \tag{5.2}$$

where τ is the accrual period between coupons. The left hand side of the equation is the spread leg of the CDO, represents the coupon paids on the non-defaulted proportion of the underlying portfolio. The right hand side is the sum of the compensation for the losses occurring between t_{i-1} and t_i, assuming that these losses are paid at the end of each accrual period. From the above we shall then focus on calculating the expected value of $f(q_T)$ under the pricing risk neutral measure.

Since conditional on the vector $\underline{\tau}$ the process q_t is normally distributed with parameters $\mu(\underline{\tau})$ and $C(\underline{\tau})$, the conditional loss process $L(T|\underline{\tau}) = 1 - \sum_i q_i(T|\underline{\tau})$ and therefore is normal distributed with mean $1 - \sum_i \mu_i(\underline{\tau})$ and variance $\sum_{i,j} C_{ij}(\underline{\tau})$. However, the following result shows that to estimate the parameters of the distribution of $L(t|\underline{\tau})$ is not required to calculate $C(\underline{\tau})$ explicitly.

Proposition 5.1. *If conditional on the vector $\underline{\tau}$ all obligors are independent with probability of default p_i given by the initial rate i at time 0, then the random variable*

$L(T|\underline{\tau})$ is normal with the following parameters

$$\tilde{p}(\underline{\tau}) = \sum_i q_i(0)p_i(\underline{\tau}) \tag{5.3}$$

$$\tilde{\sigma}^2(\underline{\tau}) = q_i(0)p_i(\underline{\tau})(1 - p_i(\underline{\tau})) \tag{5.4}$$

and the price of the equity tranche with attachment point K is given by

$$\mathbb{E}_Q\{f(t,\underline{\tau})\} = \mathbb{E}_Q\{f(t|\underline{\tau})|L(t|\underline{\tau}) \leq K/(1-R)\} + KP(L(t|\underline{\tau})(1-r) > K)$$

$$= \int_0^{K'} z(1-R)\phi\left[x; \tilde{p}, \tilde{\sigma}/\sqrt{n}\right] dz + K\Phi\left(-\frac{K' - \tilde{p}}{\tilde{\sigma}/\sqrt{n}}\right) \tag{5.5}$$

where $K' = K/(1-r)$.

Proof. Since for $n > 0$ we can express

$$\sum_i q_i^n(t) = \sum_i \sum_j \hat{q}_j^{n,i}(t)$$

where $\hat{q}^{n,i}(t)$ represents a portfolio with initial position of $q_i^n(0)$ units in rate i. Since conditional on the environment process ξ_s the elements in the portfolio are independent and homogeneous up to the credit rating, it is true that

$$\text{var}\left(\sum_i q_i^n(t)\middle|\underline{\tau}(t)\right) = \sum_i (q_i^n(0))^2 \text{var}\left(\sum_j \hat{q}_j^{1,i}(t)\middle|\underline{\tau}(t)\right)$$

$$= \sum_i (q_i^n(0))^2 p_i(1-p_i)$$

since $\sum_j \hat{q}_j^{1,i}(t) = 0$ if and only if default has occurred by t and is equal 1 otherwise, therefore is Bernoulli p_i. Now by Theorem 4.2 we know that conditional on $\underline{\tau}(t)$

$$\frac{1}{\sqrt{n}}\left(\hat{q}^{n,i}(t) - nq^{(0,i)}(t)\right) \to^d q^{(1)}(t)$$

where $q^{(0,i)}(t)$ denotes the fluid limit of $\hat{q}^{n,i}(t)$. By the Helly–Bray lemma ([2], pp. 251) we know that

$$\text{var}\left(\sum_j \hat{q}_j^{n,i}(t)\middle|\underline{\tau}(t)\right) \to \text{var}\left(q^{(1,i)}(t)\middle|\underline{\tau}(t)\right) \tag{5.6}$$

where $q^{(1,i)}(t)$ denotes the diffusion limit associated to $\hat{q}^{n,i}(t)$. Since the fluid limit is deterministic and $\sum_j \hat{q}^{n,1}$ is a sum of identical Bernoulli random variables

$$\text{var}\left(\frac{1}{\sqrt{n}}\left(\hat{q}^{n,i}(t) - nq^{(0)}(t)\right)\right) = p_i(\underline{\tau})(1 - p_i(\underline{\tau})) \tag{5.7}$$

and then (5.4) follows. Since L_t is normally distributed conditional the rest of the results is trivial. □

Therefore, when interested exclusively in the default process of the portfolio, rather than the rating distribution, it is possible to obtain the conditional distribution of the defaults from the fluid limit without solving the Lyapunov equation (4.19).

5.2. *Changes of measure, the Poisson space and Quadrature formulas*

By conditioning on the actual number of jumps in the environment process we can rewrite the pricing formula as

$$
\mathbb{E}(f(T)|\underline{\tau})) = \sum_{i=0}^{\infty} P(N_T = i)\mathbb{E}(f(T)|\underline{\tau}; N_T = i)
$$

$$
= \sum_{i=0}^{\infty} P(N_T = i)\mathbb{E}(f(T)|(\tau_1, \ldots, \tau_i))
$$

$$
\approx \sum_{i=0}^{N^*} P(N_T = i)\mathbb{E}(f(T)|(\tau_1, \ldots, \tau_i)) \tag{5.8}
$$

where N^* is set such that the rest of the series is small enough. The process N_t is a Poisson process when $\lambda_\xi = \lambda$ for all $\xi \in \Xi$ and this conditioning and integrating approach to the process is related to the canonical space of a Poisson process [17], we discuss this below. To derive the quadrature approximation, first we have to study the distribution of the jump times in the environment process. Each of the conditional expectations $F(T, \tau_1, \ldots, \tau_N) = \mathbb{E}_\mathbf{Q}(f(T)|(\tau_1, \ldots, \tau_N))$ may be calculated using Gaussian Quadratures. For simplicity of notation we will assume $T = 1$ in the following development.

We are interested in applying Gaussian Quadrature methods in the simplex set S^N to estimate integrals of the form

$$
\int_{S^N} f(\underline{\tau}) \cdot w(\underline{\tau}) d\underline{\tau}
$$

where w is the weight function (density) and f is the payoff of the claim to be priced.

The concept of change of measure is useful to calculate $W_N(A)$, the conditional probability of the set A conditional on $N_T = N$, and the probability $P(N_T = N)$. In the following consider $\mathbf{K}$ the index set for the set of all arrays of possible paths for ξ_t characterized by the sequences of states $(i_0^k, \ldots, i_N^k) \in \Xi^{N+1}$ and the number of transitions $\{n_{ij}^k\}$, $k \in \mathbf{K}$ and $\sum_{i,j} n_{ij} = N$. By assuming a change to a measure $\tilde{\mathbf{P}}$ where ξ_t follows a Markov chain with generator matrix Υ such that $\upsilon_{ij} = \overline{\upsilon}$ for $i \neq j$ and $\upsilon_{ii} = n_x \upsilon$, applying the Girsanov formula (see Sec. IV.22 in [18]) we obtain

$$P(N_T = N) = \mathbb{E}\left\{\mathbb{I}_{N_T = N}\frac{d\mathbf{P}}{d\tilde{\mathbf{P}}}\right\} \tag{5.9}$$

$$= \frac{1}{N!}\sum_{k \in \mathbf{K}}\prod_{i \neq j}v_{ij}^{n_{ij}^k}\int_{S^N}\exp\left(-\sum\Upsilon_{i_j^k}\tau_{j+1}\right)d\underline{\tau}$$

where $\tau_{j+1} = 1 - \sum^N\tau_j$. Using the same argument, for $A \in S^N$ measurable

$$W_N(A) = \mathbb{E}\left\{\mathbb{I}_A\frac{d\mathbf{P}}{d\tilde{\mathbf{P}}}\right\}$$

$$= \frac{1}{P(N_t = N)}\sum_{k \in K}\frac{1}{N!}\prod_{i \neq j}v_{ij}^{n_{ij}^k}\int_A\exp\left(-\sum v_{i_j^k}\tau_{j+1}\right)d\underline{\tau} \tag{5.10}$$

from where it is possible to derive the density function

$$w_N(\underline{\tau}) = \frac{1}{P(N_t = N)}\sum_{k \in \mathbf{K}}\prod_{i \neq j}\left(\frac{v_{ij}}{\bar{q}}\right)^{n_{ij}^k}\sum_{j=0}^N v_{i_j^k}\tau_{j+1}\exp\left(-\sum v_{i_j^k}\tau_{j+1}\right) \tag{5.11}$$

5.2.1. *The canonical space of a Poisson process*

The decomposition of the sample space of the environment process in a series of simplex sets is similar to the analysis of a Poisson process by Neveu [17] in what is known as the canonical space of a Poisson process.

It is well known that conditional in the number of jumps in a given interval of time, the inter-arrvial times of a Poisson process are uniformly distributed. Neveu defines a Poisson process in terms of the interarrival times as points uniformly distributed in n-dimensional hypercubes $n = 1, \ldots$. Moreover, he defines a point process in terms of the distribution of these points in the hypercubes. By using the pricing approach we have developed a framework similar to the one defined by Neveu taking simplex sets instead of hypercubes. Both approaches are equivalent in the case of a Poisson process, we now explain this.

The canonical space of a Poisson process in the interval $[0, T]$ is the probability space $(\hat{\Omega}, \hat{\mathcal{B}}, \hat{\mathbf{P}})$ defined as follows: The sample space $\hat{\Omega}$ is formed for all vectors $\mathbf{t} \in \mathbb{R}^n$ for $n \in \mathbb{N}$ and a "distinguished" element $[0, T]^0 = \{a\}$, i.e., $\hat{\Omega} = \bigcup_{n=0}^\infty [0, T]^n$. The σ-field is $\hat{\mathcal{B}} = \{G : G \cap [0, T]^n \in \mathcal{B}([0, T]^n)\}$ where $\mathcal{B}([0, T])^n$ denotes the Borel algebra of $[0, T]^n$ for $n \geq 1$. Finally, if μ^n is the Lebesgue measure in $[0, T]^n$ and $\mu^0 = \delta_a$, then for $A \in \hat{\mathcal{B}}$,

$$\hat{P}(A) = \sum_{i=0}^\infty e^{-\lambda}\frac{\lambda^n}{n!}\mu^n(A \cap [0, T]^n) \tag{5.12}$$

On this probability space the random variable

$$\hat{N}_t\{a\} = 0 \tag{5.13}$$

$$\hat{N}_t(t_1, \ldots, t_n) = \sum_{i=1}^{n} \mathbb{I}_{[0,t]}(t_i) \tag{5.14}$$

is a Poisson process with intensity rate λ. Notice that in the vector $\underline{t}$ the order is not important, if $\Pi(\underline{t})$ is a permutation of the elements of $\underline{t}$, $\hat{N}_t(\Pi(\underline{t})) = \hat{N}_t(\underline{t})$. Under this approach, the Poisson process is a particular case of point process. Point processes can be thought as a population process, the vector $\underline{t}$ represents the elements of $\mathbb{R}$ in the current population, where a represents the empty population. From the characteristics of the Poisson space we know that the jump times in process are uniformly distributed conditional on the number of jumps.

In the exposition of the previous sections we assume the probability space $(\Omega, \mathfrak{B}, \mathbf{P})$ as follows. The sample space $\Omega = \bigcup_{n=0}^{\infty} S^n$, where $S^n = \{\underline{t} \in \mathbb{R}^{n+} : \sum_{i=0}^{n} \leq T\}$ is a simplex in positive part of $\mathbb{R}^n$, and $S^0 = \{a\}$. The σ-algebra is $\mathfrak{B} = \{G : G \cap [0,T]^n \in \mathfrak{B}(S^n)\}$. With the measure $W_n(\cdot)$ in S^n defined as in (5.10) for $n \geq 1$, and $W_0 = \delta_a$. We define for $A \in \mathfrak{F}$,

$$P(A) = \sum_{i=0}^{\infty} P(N_T = n) W_n(A \cap S^n). \tag{5.15}$$

In this form the probability space $(\Omega, \mathfrak{B}, \mathbf{P})$ defines a point process as described by [17]. We then define the counting process

$$N_t\{a\} = 0 \tag{5.16}$$

$$N_t(\tau_1, \ldots, \tau_n) = \sum_{i=1}^{n} \mathbb{I}_{[0,t]}(\tau_1 + \cdots + \tau_i) \tag{5.17}$$

In the case that $\lambda_i = \lambda$ for all i, the measure $W_n = n!\mu^n$ where μ^n is the Lebesgue measure in the simplex S^n and μ^0 is as above. In this case the probability measure (5.15) is

$$P(A) = \sum_{i=0}^{\infty} e^{-\lambda_0} \frac{\lambda_0^n}{n!} n! \mu^n (A \cap S^n) \tag{5.18}$$

and N_t is an standard Poisson process. For some n consider $A \in S^n$ a measurable set (with respect to $\mathfrak{B}$), then the set $\hat{A} = \{\Pi(\underline{t}) : \underline{\tau}(\underline{t}) \in A\}$ is a measurable set with respect to $\hat{\mathfrak{B}}$. Notice that for almost all points (with respect to the Lebesgue measure) in S^n the number of permutations of the vector $\underline{t}(\tau)$ is $n!$, therefore $W_n(A) = \hat{\mu}(\hat{A})$. Finally notice that $\hat{N}_t(\underline{t}) = N_t(\underline{\tau})$ for all $t \in [0,T]$ if and only if $\underline{t} = \Pi(\underline{t}(\underline{\tau}))$ for some permutation, from this we conclude that N_t is a Poisson process with intensity λ.

In the following section we will present numerical integration techniques and we will discuss its implementation in the space $(\Omega, \mathfrak{B}, \mathbf{P})$ with a sample space as the union of simplex sets. In the case of the environment process being a Poisson process with intensity λ it does not make any difference the Poisson space we use and we have no special reason to prefer the use of our proposed space (of simplex sets).

It is always possible to translate the problem to a Poisson type by using the Girsanov theorem for Markov chains. In this case we have a constant weight function (if the Radon–Nikodym is not accounted in the weight function) and then the use of the canonical space may be preferred.

5.2.2. *Gaussian quadratures*

Quadrature methods for numerical integration have been widely used as an alternative to the traditional *Newton–Cotes* approach. The appealing feature of quadratures is the lower number of points of evaluation required to obtain a desired degree of accuracy defined in terms of the maximum degree of the polynomials for which the formula is exact.

For example, on the real line a Gaussian quadrature of degree m, with respect to a the defined measure μ, provides an exact formula for polynomials up to degree $2m - 1$, this is

$$\int_{\mathbb{R}^n} x_i^k d\mu = \sum_{j=1}^{m} w_j x_{j,i}^k \qquad (5.19)$$

where $x_i, i = 1, \ldots, n$, is the ith component of x and $k \leq 2m - 1$. This result is the fundamental theorem of Gaussian Quadratures (see [19]). While the property of the formula being exact up to double degree polynomials is appealing, the selection of the abscissas x_j is not always trivial. The most common Gaussian Quadrature is the one related to the Lebesgue measure (know as Gaussian–Lebesgue). Since the the density function involved in our calculations has the exponential form we will prefer the use of the Gaussian–Lebesgue quadratures over other less stable and more expensive methods as proposed in [14].

For the multiple dimensional integration problem, we can express formulas as a "product" of one dimensional quadratures, therefore these cases are referred as *product formulas*. A survey of these methods can be found in [20]. We will just explain briefly the case of a simplex set since it will be useful to our pricing technique.

Consider an integral of the form

$$I = \int_0^1 \int_0^{1-x_1} \cdots \int_0^{1-x_1-\cdots-x_n} f(x_1, \ldots, x_n) dx_n \cdots dx_1, \qquad (5.20)$$

setting the change of variable

$$x_1 = y_1 \qquad (5.21a)$$

$$x_2 = y_2(1 - y_1) \qquad (5.21b)$$

$$\vdots$$

$$x_n = y_n(1 - y_1) \cdots (1 - y_{n-1}) \qquad (5.21c)$$

we obtain

$$I = \int_{[0,1]^n} f(x_1(y_1), \ldots, x_n(y_n))(1 - y_1)^n (1 - y_2)^{n-1} \cdots (1 - y_{n-1}) dy_n \cdots dy_1$$

If a quadrature formula $\{w_i, x_i\}_{i \in I}$ of degree m can be derived in the real interval $[0, 1]$, then $\{\prod_{j=1}^n w_{i_j}, (x_{i_1}, \ldots, x_{i_n})\}$ defines a quadrature formula in $[0, 1]^n$ that is exact for all polynomials with coefficients less or equal to m. Here $\{(i_1, \ldots, i_j)\}$ represents the set of all possible combinations of indices $i \in I$. From the change of variable above, the integral in the simplex is taken to the Cartesian product of the interval $[0, 1]$ and product formulas can be applied.

It is evident that product formulas lose their appeal as the dimension increases. As an alternative there are available in the literature some other quadrature formulas. In particular we use the method proposed by Grundmann and Möller [9] that claims to use the minimum number of integration nodes.

5.3. *Some comparisons*

We present an example to illustrate the advantages of Gaussian Quadratures and the use of change of measures. We set a two state model with two rates, the following transition rates (corresponding to each environment state)

$$T_0 = \begin{pmatrix} 0.0125 & 0.0200 \\ 0.0100 & 0.0375 \end{pmatrix} \qquad T_1 = \begin{pmatrix} 1.500 & 0.030 \\ 0.005 & 3.000 \end{pmatrix}$$

and environment process jump rates

$$\lambda = \begin{pmatrix} 0.05 \\ 1.0 \end{pmatrix}$$

We evaluate the expected loss in different equity tranches using different methods. Our aim is to achieve 6-digit accurate estimation, this is particularly difficult under Monte Carlo methods, particularly for large tranches. To complete our initial information, we assume a portfolio with initial distribution 50/50, 100 elements and a recovery rate $r = 40\%$.

First, we implement Monte Carlo simulation with $1,000,000$ simulations, and estimate the mean and variance of the loss in each tranche. The results are shown in Table 1. By (4.22) in Proposition 4.2 we find that the exact probability of default is 0.047207. From the results obtained it can be seen that in order to achieve a 6-digit (95% confidence) estimation in the 22% tranche we need over $1,199$ million simulations.

We now present some results (Table 2) using importance sampling techniques. For example, in the case we change the measure such that the jump rate λ_0 increases to 0.5 (second column in the table) the standard deviation observed in the tranche 0–6% is 81bps equivalent to a reduction of 14% with respect to the original measure. There are two parts that require different strategies, first, the equity part where both events (low and high default rate states) define the expected loss in the tranche.

Table 1. Monte Carlo results.

Attachment Point	Expectation (%)	Std. Dev. (bps)	95% Confidence (%) Interval ($\pm$)
3	1.5343	32	6.40E-04
6	1.6919	94	1.85E-03
9	1.8240	155	3.04E-03
12	1.9493	213	4.18E-03
22	2.3145	390	7.64E-03
prob. Def.	4.7240	1131	2.22E-02

Table 2. Importance sampling: Standard deviation (bps).

Base Tranche	New Rates λ_0/λ_1			
	0.05/1.0	0.5/1.0	0.2/2	0.1/2
3%	32	96	40	24
6%	94	81	37	64
9%	155	66	60	112
12%	213	55	91	160
22%	390	52	196	312
Percentage Reduction				
3%				25%
6%		14%	61%	32%
9%		57%	61%	27%
12%		74%	57%	25%
22%		87%	50%	20%

However this already shows relatively low variance. Secondly, the senior tranches will be affected in the the case of high default rates.

We must consider that the exponential form of the Radon–Nikodym derivative causes high variance when the jump rates are increased dramatically. Therefore while it would seem ideal to increase the rate of occurence of jumps to eliminate the variance, the Radon–Nikodym derivative form makes this not practical.

While the reduction in the senior tranches is outstanding when we increase the jump rate to the stressed economic scenario, the number of simulations required still being relatively high.

In order to implement the Gauss quadratures we analyze the probability distribution of the number of jumps in the environment process. From Table 3 it is clear that by conditioning in the number of jumps, just a small number of conditional expectations will be enough for an accurate valuation. We will use 6 jumps, however it can be easily work with just four (the probability of a 5 jumps is lower than the order of accuracy and the expected loss is less than 1). Furthermore, the error tolerance required decreases while the dimension (number of jumps) increases since the probability of higher number of jumps is smaller.

Table 3. Probability distribution of jumps.

n	Probability	Accumulated
0	0.951229	0.951229
1	0.030703	0.981932
2	0.017746	0.999678
3	0.000252	0.999930
4	6.90E-05	0.999999
5	6.26E-07	1.000000
6	1.11E-07	1.000000

First, we implement Gaussian–Legendre quadratures extended to a simplex as explained in Sec. 5.2. The second method uses ad hoc quadratures extended to a simplex; for lower order degree we derive directly the orthogonal polynomials and coefficients for the Jacobi Matrix, larger order quadratures use the sequence approach of [14]. A third approach is based on a change of measure to a homogeneous rate Poisson and applying the Gaussian–Legendre quadrature that is suitable since the conditional probability is uniform. We try two changes of measure: one to a Poisson with rate 0.25 that keeps a stable Radon–Nikodym derivative, and a second to a Poisson rate 0.05 where the probability of higher number of jumps vanishes faster. Finally, we implement a 5-degree Grundmann–Möller quadrature for a simplex presented in [9].

In Table 4 we can observe the number of points required to match a 6-digit accurate estimation in the loss of the tranche. The first conclusion is that in all cases it is clear the advantage with respect to Monte Carlo methods. Even when producing quadrature points is more expensive than Monte Carlo points, the incredible reduction of points required justifies the use of quadratures.

We compare in Table 5 the most accurate result (Obtained with a Gaussian Legendre quadrature using 224 983 points) with our Monte Carlo simulation and a 5 degree quadrature ad hoc for simplex domains developed in [9] that claims to have the minimum number of integration nodes.

Table 4. Points required for 6-digit accuracy.

Method	Attachment Point					
	3	6	9	12	22	Total
Gauss-Legendre	232	332	445	657	633	467
Gauss-Exponential	232	361	417	837	662	1023
Change of Measure (1)	212	236	606	657	625	225
Change of Measure (2)	433	437	595	595	595	341
Grundmann Moller (*)	127	127	127	127	127	127

Note: (*)Number of point associated to Grundmann Möller degree 5 regardless of accuracy.

Table 5. Expected loss (%).

Tranche	Gauss Quadrature	Monte Carlo (1 mm)	Grundmann Möller $d = 5$
3	1.5344	1.5343	1.5351
6	1.6921	1.6919	1.6965
9	1.8243	1.8240	1.8356
12	1.9496	1.9493	1.9705
22	2.3145	2.3145	2.2919
def prob	4.7207	4.7240	4.7212

6. Calibration

We consider the pool of CDS in the iTraxx Europe index for quotes from 11 April to 15 April 2005 and from 26 February to 5 March 2007. This data is presented in Tables 7 and 8, and gives the par spreads for 5-year protection on the tranches specified in the left-hand column. As is customary, the premium payment for the equity (0–3%) tranche is an up-front payment of a percentage (this is the number given in the table) of the notional amount, plus a running spread of 500bp. For example, the figures given for April 11 in Table 7 mean that equity tranche investors pay 25% up-front plus 500bp, while investors in the 3–6% tranche pay nothing up-front but a running spread of 161bp (and similarly for all other tranches). The calibration is performed by minimizing the square error of the pricing error (spreads) for the CDO iTraxx quotes and the CDS quote. The minimization routine is performed combining a gradient based method and a pattern search method from Matlab. The reason to alternate methods is to avoid local minima (by using pattern search) while keeping speed (gradient method).

First, we performed the calibration assuming a 2-state environment process. As for the rating process, we assume both a single and a two-rate portfolio.

The calibration of the two-state model to the market data is poor. It is possible to improve the fitting of the senior tranche and represent a increasing behavior in the base correlation by increasing the default rate μ_1 in the catastrophic state and diminishing its probability of occurrence by decreasing λ_0, however this increases the error in the mezzanine tranche prices.

Various examples show us that the tranche prices generated by a two-rate system can be accurately matched by a model using only a single rate. This may be explained by the fact that in a large portfolio the effect of idiosyncratic risk events is averaged inside the portfolio, by using the fluid and diffusion approximation we are then dampening this effect. On the other side, the systematic risk (represented by the random environment process) is the main driver of the model.

6.1. *A 3-state environment process*

The high level of the implied and base correlations associated to the senior tranches can be explained as the premium asked for the existence of a rare catastrophic event where a high proportion of default occur. We introduce a third environment state such that a two state switching regime provides the default structure of the equity and junior mezzanine tranches and the third state provides the "worst scenario" that adds value to the senior tranches.

6.1.1. *Implementation*

The valuation methodology for the 3-state model is the same as explained in the previous section. In addition to the transition matrices A_1, A_2 and A_3 a generator matrix $Q \in \mathbb{R}^{3 \times 3}$ for the environment process is assumed. Default probabilities can be calculated using Proposition 4.2.

To use a quadrature approach for general claims we must consider all possible paths in the environment process. In the general case of n-states we have $(n-1)^k$ possible paths with $k-1$ jumps. In the case of Q as described above and considering up to 6 jumps, we must consider 127 paths. We decide to use product formulas from Lebesgue–Gaussian quadratures for the paths up to 3 jumps and the Grundmann-Möller quadrature formula of fifth degree thereafter in order to reduce the number of points valuated.

We assume a model with a single rating class, a 3-state environment process with transition matrix Q, and a 3-vector $\mu = (\mu_1, \mu_2, \mu_3)$ of default rates in the 3 environment states, giving a total of 9 parameters (Q is specified by 6 parameters since the row sums are zero). We employ a combination of gradient based algorithm and pattern search algorithms provided by Matlab. In order to increase the speed we generate a common set of nodes for the quadratures according to a pre-specified distribution and update the Radon-Nikodym derivative as we modify the parameters. This method seems to improve the performance without losing accuracy. A more refined method must consider readjusting the measure of reference periodically. Applying this method to the 11 April 2005 data we obtain the following parameters:

$$Q = \begin{pmatrix} -0.080931 & 0.076783 & 0.004148 \\ 3.204384 & -3.239922 & 0.035538 \\ 4.935346 & 5.732706 & -10.668052 \end{pmatrix}$$

$$\mu = (\,0.00365 \quad 0.09194 \quad 3.73706\,)$$

By 2007 credit spreads had dropped by almost half, this could be explained by a change in the "steady" default rate or a change of appreciation of the catastrophic state. The new parameters for 27 February 2007 are

$$Q = \begin{pmatrix} -0.244905 & 0.238673 & 0.006232 \\ 2.908366 & -2.959641 & 0.051275 \\ 2.227186 & 0.269876 & -2.497042 \end{pmatrix}$$

$$\mu = (\,0.00253 \quad 0.01174 \quad 0.19825\,)$$

Table 6. Calibration error.

Tranche	27-Feb-07			11-Apr-05		
	Market	Model	Error	Market	Model	Error
0–3% (%)	8.661	8.661	0.000	25.000	25.000	0.000
3–6% (bps)	44.025	44.032	0.007	161.000	161.000	0.000
6–9% (bps)	11.875	11.877	0.002	51.500	51.500	0.000
9–12% (bps)	5.691	5.695	0.003	24.000	24.000	0.000
12–22% (bps)	2.041	2.034	−0.007	14.000	14.000	0.000
CDS (bps)	22.100	22.098	−0.002	40.000	40.000	0.000

Table 7. Calibration to iTraxx mid market quotes (April 2005).

	11-Apr	12-Apr	13-Apr	14-Apr	15-Apr
0–3%	25.00	23.95	23.95	26.75	30.63
3–6%	161.00	153.00	151.50	169.50	190.00
6–9%	51.50	47.75	47.00	53.50	61.00
9–12%	24.00	22.50	22.50	26.00	30.00
12–22%	14.00	13.00	13.00	14.50	16.25
CDS	40.00	39.00	39.00	42.00	45.00
μ_1	0.3656	0.3539	0.3543	0.3853	0.4252
μ_2	9.1943	8.9888	8.8852	8.9338	8.6264
μ_3	373.7	360.1	359.8	351.6	279.2
q_{12}	7.6784	8.5438	8.7701	7.4639	7.7863
q_{13}	0.4148	0.4565	0.4722	0.3912	0.6284
q_{21}	320.4	356.3	363.9	290.1	296.1
q_{23}	3.5538	1.8432	1.2611	7.2539	2.4838
q_{31}	493.5	491.5	492.5	488.7	515.7
q_{32}	573.3	594.1	594.3	576.1	472.7

Table 8. Calibration to iTraxx mid market quotes (Feb–Mar 2007).

	27-Feb	28-Feb	1-Mar	2-Mar	5-Mar
0–3%	8.66	9.95	10.22	11.02	12.73
3–6%	44.03	45.26	47.81	49.16	53.08
6–9%	11.88	12.52	13.01	13.65	14.94
9–12%	5.69	5.45	5.73	6.10	6.95
12–22%	2.04	2.13	2.21	2.39	2.56
CDS	22.10	22.88	23.19	23.78	25.05
μ_1	0.2527	0.2927	0.2865	0.2814	0.2984
μ_2	1.1174	0.7726	0.8486	1.0725	1.1843
μ_3	19.8249	20.4257	20.5219	20.1073	19.2929
q_{12}	23.8673	22.8874	25.0613	23.2297	21.1600
q_{13}	0.6232	0.7383	0.7352	0.6961	0.6864
q_{21}	290.8	283.0	280.6	282.3	294.4
q_{23}	5.1275	3.6429	4.0824	5.2553	8.1463
q_{31}	222.7	217.7	228.5	219.3	215.3
q_{32}	27.0	40.2	27.6	32.6	38.0

We can observe that while the model indicates a reduction in the prevalent instantaneous default rate, the default rates associated to the stressed scenarios have been brought down significatively. Table 6 shows the calibration error. Essentially, exact calibration is achieved across all tranches, supporting the argument of the existence of a third catastrophic state to explain the implied correlation structure observed in the market.

Finally, Tables 7 and 8 show the calibrated parameters for the weeks beginning 11-Apr-05 and 27-Feb-07 respectively. In both cases the model shows stability in the parameters despite the fact that the sample week from 2005 is particularly volatile.

7. Conclusions

We have presented a credit risk model whose interpretation and dynamic are intuitive. This model is intended for use in large portfolio modelling, CDO pricing and risk management. Based on well known results from stochastic networks theory, the model considers the systemic risk of default by postulating the existence of a exogenous environment state (that can be interpreted as the economic cycle). It is assumed that in a large portfolio the idiosyncratic risk (represented by the rating system) is averaged in the portfolio.

We provide a description of the asymptotic behavior of the model and use it to approximate the value of claims written in a large portfolio such as CDOs.

For pricing purposes we make use of Quadratures applied to a decomposition to the sample space similar to the characterization of point processes in Neveu [17]. We show that this technique gives us an algorithm that is fast and accurate to calculate the distribution of losses in the two-state model.

We present some numerical results that show the ability of the model to represent the skew present in the market in terms of base correlation for different periods of time.

References

[1] G. L. Choudhury, A. Mandelbaum, M. I. Reiman and W. Whitt, Fluid and diffusion limits for queues in slowly changing environments, *Comm. Statist. Stochastic Models* **13**(1) (1997) 121–146.

[2] Y. S. Chow and H. Teicher, *Probability Theory*, Third Edition (Springer-Verlag, New York, 1997).

[3] J. Christensen, E. Hansen and D. Lando, Confidence sets for continuous-time rating transition probabilities, *Journal of Banking and Finance* **28**(11) (2004) 2575–2602.

[4] *CUSP and PortfolioRisk+* (Credit Suisse First Boston, London, 2005).

[5] J. C. Esparragoza-Rodriguez, Credit risk modelling. PhD Dissertation, Imperial College, University of London (2007).

[6] S. N. Ethier and T. G. Kurtz. *Markov Processes* (John Wiley & Sons Inc., New York, 1986).

[7] H. Frydman and T. Schuermann, Credit rating dynamics and markov mixture models, Technical Report 04-15, Wharton Financial Institutions Center (2004).

[8] G. Giampieri, M. Davis and M. Crowder, Analysis of default data using hidden Markov models, *Quant. Finance* **5**(1) (2005) 27–34.

[9] H. Grundmann and H. Möller, Invariant integration formulas for the n-simplex by combinatorial methods, *SIAM Journal on Numerical Analysis* **15**(2) (1978) 282–290.

[10] *CreditMetrics — Technical Document* (J.P. Morgan & Co. Incorporated, New York, 1997).

[11] D. L. Iglehart and W. Whitt, Multiple channel queues in heavy traffic. I, *Advances in Appl. Probability* **2** (1970) 150–177.

[12] R. A. Jarrow, D. Lando and F. Yu, Default risk and diversification: Theory and empirical implications, *Math. Finance* **15**(1) (2005) 1–26.

[13] T. G. Kurtz, Strong approximation theorems for density dependent Markov chains, *Stochastic Processes Appl.* **6**(3) (1977/78) 223–240.

[14] R. Mach, Orthogonal polynomials with exponential weight function in a finite interval and application to the optical model, *Journal of Mathematical Physics* **25**(7) (1984) 2186–2193.

[15] A. Mandelbaum, W. A. Massey and M. I. Reiman, Strong approximations for Markovian service networks, *Queueing Systems Theory Appl.* **30**(1–2) (1998) 149–201.

[16] L. McGinty, R. Ahluwalia, E. Beinstein and M. Watts, *Introducing Base Correlations* (JPMorgan, 2004).

[17] J. Neveu, Processus ponctuels, *Ecóle d'Eté de Probabilites de Saint-Flour VI* (1976) 249–305.

[18] L. Rogers and D. Williams, *Diffusions, Markov Processes and Martingales*, Volume II. (Cambridge University Press, Cambridge, 2000).

[19] A. H. Stroud, *Approximate Calculation of Multiple Integrals* (Prentice-Hall Inc., Englewood Cliffs, N.J., 1971).

[20] A. H. Stroud and D. Secrest, *Gaussian Quadrature Formulas* (Prentice-Hall Inc., Englewood Cliffs, N.J., 1966).

[21] W. Whitt, *Stochastic Process Limits: An Introduction to Stochastic Process Limits and their Application to Queues* (Springer-Verlag, New York, 2002).

EMPIRICAL COPULAS FOR CDO TRANCHE PRICING USING RELATIVE ENTROPY

MICHAEL A. H. DEMPSTER*, ELENA A. MEDOVA[†] and
SEUNG W. YANG[‡]

Centre for Financial Research
Judge Business School
University of Cambridge
United Kingdom

and

Cambridge Systems Associates Limited
Cambridge, United Kingdom
*mahd2@cam.ac.uk.
[†]eam28@cam.ac.uk.
[‡]swy21@cam.ac.uk.

We discuss the general optimization problem of choosing a copula with minimum entropy relative to a specified copula and a computationally intensive procedure to solve its dual. These techniques are applied to constructing an empirical copula for CDO tranche pricing. The empirical copula is chosen to be as close as possible to the industry standard Gaussian copula while ensuring a close fit to market tranche quotes. We find that the empirical copula performs noticeably better than the base correlation approach in pricing non-standard tranches and that the market view of default dependence is influenced by maturity.

Keywords: Portfolio credit risk; CDO; copula; entropy; non-parametric estimation.

1. Introduction

Copula methods for pricing *collateral debt obligations* (CDOs) in general assume some parametric form for the copula of default times and try to obtain the values for model parameters which produce prices that most closely match those of the market. An unsatisfactory aspect of these methods is that they offer little underlying rationale for copula choice. This paper shows how to choose the copula empirically by optimizing its entropy. The strength of the entropic approach is that it provides an information-theoretic rationale for the choice of the copula and also results in

---

*Corresponding author.

excellent fits to data. By minimizing the relative entropy with respect to the industry standard Gaussian distribution, we choose the copula that is closest to the standard while ensuring a close fit to market prices.

Our method is similar in spirit to Hull and White [13], who imply from market data an empirical copula in the standard one-factor framework using the criterion of maximal smoothness. We essentially follow the same methodology as theirs but in a more general framework using the criterion of minimum relative entropy. Hull and White's method is limited to calibration to single-maturity data, and is not easily extendable to non-constant hazard rates. The entropic copula approach, however, can be used to calibrate to data across different maturities and naturally accommodates any stochastic hazard rate model. Both these methods promise perfect to near-perfect fits to the data.

The remainder of this introduction discusses credit risk modeling and CDO tranches. Section 2 describes the principle of minimum relative entropy and how to express it numerically and Sec. 3 discusses the maximum entropy copula problem. This is extended to the minimum relative entropy copula problem for CDO tranche pricing in Sec. 4, where computational results based on market data are presented. Section 5 concludes.

1.1. *Correlated intensities in portfolio credit risk modeling*

In *single-name* credit risk modeling, there are two main approaches: the *structural approach* and the *reduced form approach*. The latter has been more popular in pricing applications because it generally offers better fits. A typical example of the reduced form approach is to assume that default occurs when a doubly-stochastic Poisson process (also called a Cox process) first makes a jump.

Extending credit risk modeling to *multiple names* introduces an extra complication. The interdependence between firms in their probability of default is an important aspect that must be taken into account. Some early reduced form approaches attempted to model this dependence by allowing the stochastic intensities of the Cox processes to be correlated and the default events conditioned on the intensities to be independent. Several examples of these models can achieve relatively close fits to market data, for example Mortensen [18] and Graziano and Rogers [11]. As Mortensen [18] conceded, however, the resulting model parameters can be unrealistic because an unnaturally high degree of correlation between the intensities is needed to reproduce the observed market prices. Moreover, Das *et al.* [6] in their empirical study concluded that the level of default dependence that can be realistically introduced by this technique is not sufficient to capture the clustering of defaults that are observed in the market. Why this is the case can be appreciated when we remember that the probabilities of default we are dealing with are very low. To achieve significant clustering of defaults, the default probabilities must be wildly fluctuating at unrealistic levels, as well as being highly correlated.

1.2. *Copulas*

The most popular method for modeling *portfolio credit risk* has been to use copulas. Copulas are used to introduce dependence between default times in a direct way, not indirectly through default intensities. This allows us to reproduce the level of clustering of defaults that we observe in reality. However, as mentioned previously, the choice of copula is rather arbitrary, motivated by two main criteria: the quality of final fit to the data, and computational tractability. As we shall see, choosing an empirical copula using the entropic approach gives us an underlying rationale for this choice.

1.3. *CDO tranche pricing*

A CDO is a derivative structure which provides protection against the loss on a portfolio of defaultable assets. The seller of protection on a tranche of this portfolio receives regular premium payments. In return, he must pay the buyer of protection any losses on that tranche that are incurred through defaults. Each tranche covers only a portion of the total potential losses of the portfolio.

To illustrate, consider a CDO portfolio of n *names*, each with *unit nominal* amount and with *maturity* T. Then denoting τ_j as the *time of default of name* j the amount lost on the portfolio at time t is

$$L(t, \boldsymbol{\tau}) = (1 - R) \sum_{j=1}^{n} 1_{\{\tau_j < t\}}, \tag{1.1}$$

where R is the constant *recovery rate* and $\boldsymbol{\tau} = (\tau_1, \ldots, \tau_n)$ is the vector of *default times*.[1] Now consider a *tranche* of this CDO with *attachment point* α and *detachment point* β. Then the *loss* on this tranche at time t will be the non-decreasing function

$$M(t, \boldsymbol{\tau}) = \begin{cases} 0 & \text{if } L(t, \boldsymbol{\tau}) \le \alpha, \\ L(t, \boldsymbol{\tau}) - \alpha & \text{if } \alpha < L(t, \boldsymbol{\tau}) \le \beta, \\ \beta - \alpha & \text{if } L(t, \boldsymbol{\tau}) > \beta. \end{cases}$$

$$= \beta - \alpha - (\beta - L(t, \boldsymbol{\tau}))^+ + (\alpha - L(t, \boldsymbol{\tau}))^+,$$

where $(x)^+ := \max\{x, 0\}$. Thus the loss on an (α, β) tranche can be written in terms of a *put spread* as shown in Fig. 1 below. Then the *default leg* — the present value of the (random) amount that the *seller* of protection needs to pay — can be written as

$$D(\boldsymbol{\tau}, T) = \sum_{j=1}^{n} e^{-r\tau_j} 1_{\{\tau_j \le T\}} \left(M(\tau_j, \boldsymbol{\tau}) - M(\tau_j^-, \boldsymbol{\tau}) \right) \tag{1.2}$$

---

[1]We use boldface throughout to denote random entities.

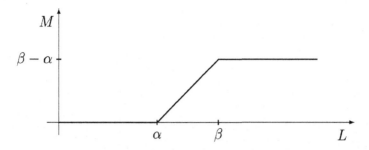

Fig. 1. Loss on a (α, β) tranche vs portfolio loss.

and the *premium leg* — the present value of the (random) amount that the seller of protection receives in return — can be written as

$$P(\boldsymbol{\tau}, T) = s \sum_{k=1}^{P} e^{-rt_k} \left(\beta - \alpha - M(t_k, \boldsymbol{\tau}) \right), \qquad (1.3)$$

where s is the premium rate, and where $t_k \leq T, k = 1, \ldots, P$, are the premium payment dates.[2]

To price a CDO tranche, we wish to work out the fair *premium rate* $\bar{s}$. Like any other swap, this is the value that makes the expected payoffs of the default and premium legs equal. Therefore the fair premium rate is the value $\bar{s}$ which satisfies

$$E[D(\boldsymbol{\tau}, T)] - \bar{s}E[P(\boldsymbol{\tau}, T)] = 0. \qquad (1.4)$$

It should be clear that it is the "tranching" feature which makes the fair premium rate $\bar{s}$ depend on the *default dependence*. If we consider a "tranche" that spans the whole portfolio, i.e., a $(0, L_{\max})$ tranche, then we would have $M \equiv L$; since L is linear in each of the default times, the expected payoffs $E[D(\boldsymbol{\tau}, T)]$ and $E[P(\boldsymbol{\tau}, T)]$ only depend on the mean of $\boldsymbol{\tau}$ and not its higher moments, and hence the fair premium rate $\bar{s}$ does not depend on the default dependence. This is not the case if the tranche only spans a subset of the overall portfolio loss.

2. Minimum Relative Entropy

The principle of minimum relative entropy is closely related to the principle of maximum entropy and it is instructive to consider the latter first.

2.1. *Principle of maximum entropy*

The *principle of maximum entropy* (MaxEnt) is a method of obtaining a unique probability distribution for a random variable from a given set of data assumed to be generated by it. The principle was first formulated by Jaynes [15] and is used in a wide variety of applied sciences.

---

[2]We ignore accrued payments here for expositional simplicity but include them in our numerical studies.

To illustrate a typical problem in finance, suppose we have a finite set of vanilla European option prices on a stock. The price of each option is a function of the risk-neutral density of the stock price at maturity. The aim of MaxEnt in this example is to infer from the given set of option prices the risk-neutral density of the stock price.

However, such a problem is generally highly under-determined because we have many fewer option prices than possible stock prices. But a unique density can be obtained if we optimize some objective function that depends on the density, while satisfying the observed market prices. Various forms for the objective function have been proposed, such as Fisher information [9, 12], maximal smoothness [13, 14] and entropy [4].

Since entropy is a measure of uncertainty, inferring the probability distribution of a random variable by maximizing entropy is optimal in the sense that we only take into account information that is given and do not assume anything else about the distribution. We choose a distribution which is consistent with the given information but otherwise has maximum uncertainty. MaxEnt applied to finance is thus related to the concept of *market efficiency* in the sense that prices fully reflect all available information in the market.

The MaxEnt principle is a *non-parametric* method of estimating a probability distribution. In *parametric* estimation the focus is on obtaining the best estimator $\hat{\theta}$ of a given parametric family $f(\cdot|\theta)$ of densities. This process involves two steps: model *specification* and model *estimation*. In non-parametric estimation the focus is on obtaining a good estimate of f *directly* from the data, which eliminates the need for model specification.

The MaxEnt principle is thus well-suited to the estimation of copulas in portfolio credit risk modelling. As discussed earlier, the choice of copula for CDO tranche pricing in much of the literature is rather arbitrary, motivated by quality of fit. These methods are parametric in nature. By using a non-parametric approach such as MaxEnt the problem of arbitrarily choosing a copula is obviated.

To set notation, suppose we observe the data set $\{(\tilde{a}_1, \bar{a}_1), \ldots, (\tilde{a}_m, \bar{a}_m)\}$ with $\tilde{a}_i, \bar{a}_i \in \mathbb{R}$ for each $i = 1, \ldots, m$. We know from the problem at hand that $E[a_i(\mathbf{x}, \bar{a}_i)] = \tilde{a}_i$ for each i, where $\mathbf{x}$ is a random vector taking values in some domain $D \subseteq \mathbb{R}^n$.

Letting f be the density for $\mathbf{x}$ (or probability mass function if discrete) we wish to maximize its *differential entropy*, i.e., solve the problem

$$\sup_{f \in L_1^+(D)} - \int_D f(x) \log f(x) dx, \tag{2.1}$$

subject to the data constraints

$$\int_D a_i(x, \bar{a}_i) f(x) dx = \tilde{a}_i \quad i = 1, \ldots, m, \tag{2.2}$$

where $a_i(\cdot, \bar{a}_i)$ is a piecewise continuous function on D and $L_1^+(D)$ denotes the non-negative cone of the space of real-valued integrable functions on D with the usual integral norm.

2.2. *Solution to the MaxEnt problem*

The Lagrangian for this problem can be easily maximized by elementary calculus of variations.[3,4] After normalizing f to make it a probability density the solution is of the form

$$\hat{f}(x, \lambda) = \frac{1}{Z(\lambda)} \exp\left\{\sum_{i=1}^{m} \lambda_i a_i(x, \bar{a}_i)\right\}, \tag{2.3}$$

where $\lambda = (\lambda_1, \ldots, \lambda_m)$ and $Z(\lambda) := \int \exp\{\sum_i \lambda_i a_i(x, \bar{a}_i)\}dx$ is the normalizing constant. The optimal f can then found by solving for the unique values λ_i that satisfy the constraints (2.2).

However, instead of solving the m simultaneous equations (2.2) with $f := \hat{f}$ to determine the lambdas, we can more elegantly solve the dual problem. Consider the dual function

$$\mathcal{L}^*(\lambda) = -\int_D \hat{f}(x) \log \hat{f}(x)dx + \int_D \lambda' \left(A(x) - \tilde{a}\right) \hat{f}(x, \lambda)dx \tag{2.4}$$

$$= \log Z(\lambda) - \lambda'\tilde{a}, \tag{2.5}$$

where $A(x) := (a_1(x, \bar{a}_1), \ldots, a_m(x, \bar{a}_m))'$ and $\tilde{a} = (\tilde{a}_1, \ldots, \tilde{a}_m)'$ and the prime symbol denotes transpose. The *dual problem* of the MaxEnt problem given by (2.1) and (2.2) is

$$\inf_{\lambda \in \mathbb{R}^m} \mathcal{L}^*(\lambda).$$

This is a much easier optimization problem than the original primal problem because it is finite dimensional and unconstrained, and we know that the dual function is always convex. Moreover, we can easily show that it is *strictly convex* if the functions a_i are linearly independent.

Proposition 2.1. *The dual function (2.4) is convex, and it is strictly convex if and only if the functions a_i are linearly independent.*

Proof. Take $\bar{\lambda}$ and $\tilde{\lambda}$ in $\mathbb{R}^m$ and set $\lambda := s\bar{\lambda} + (1-s)\tilde{\lambda}, s \in [0, 1]$. Then by Hölder's inequality

$$\mathcal{L}^*(\lambda) = \log \int \exp\left\{s\bar{\lambda}'A(x) + (1-s)\tilde{\lambda}'A(x)\right\} dx - s\bar{\lambda}'\tilde{a} - (1-s)\tilde{\lambda}'\tilde{a}$$

$$\leq s\left(\log \int e^{\bar{\lambda}'A(x)}dx - \bar{\lambda}'\tilde{a}\right) + (1-s)\left(\log \int e^{\tilde{\lambda}'A(x)}dx - \tilde{\lambda}'\tilde{a}\right).$$

If the a_i's are not linearly independent then we can find a $\bar{\lambda}$ and $\tilde{\lambda}$ with $\bar{\lambda} \neq \tilde{\lambda}$ such that $\bar{\lambda}'A \equiv \tilde{\lambda}'A$, whence Hölder's inequality becomes equality. But if the a_i's are linearly independent, then seeing Hölder's inequality as an application of Jensen's

---

[3] To apply the Euler–Lagrange equation, i.e., equate the first variation to zero, we need to assume that there exists a feasible f with finite entropy and that $f > 0$ almost everywhere.
[4] Note that the Lagrangian is concave in f.

inequality for the strictly concave function given by $f(x) := x^s$ for $s \in (0, 1)$, the result follows.

Alternatively we could show that the Hessian is the covariance matrix of the a_i's, and thus the dual function is strictly convex if the a_i's are linearly independent. □

The minimization can be solved numerically using any gradient-based optimization method such as the BFGS quasi-Newton algorithm.[5] We just need the gradient vector of the dual, whose ith element is given by

$$\frac{\partial \mathcal{L}^*(\lambda)}{\partial \lambda_i} = \int_D a_i(x, \bar{a}_i) \hat{f}(x, \lambda) dx - \tilde{a}_i.$$

We can differentiate the dual (2.4) under the integral sign since the integrand is in $L_1(D)$ for each fixed λ, and is differentiable with respect to λ for almost all x with bounded derivative for all bounded λ.

Notice that the gradient corresponds to the constraints (2.2) and that the constraints will be satisfied when the gradient vector is zero.

In addition to the assumption that the feasible set is not empty, we also impose the assumption mentioned previously in footnote 3 that there must exist a feasible f with $f > 0$ almost everywhere which has finite entropy.

2.3. *Regularization*

The Slater constraint qualification that we have just mentioned is difficult to check in general. In particular, it is difficult to determine whether the observed data set $\{(\tilde{a}_1, \bar{a}_1), \ldots, (\tilde{a}_m, \bar{a}_m)\}$ is consistent.[6] A related problem is that in real applications there may be measurement errors in the observed data — not only are these errors a problem in themselves, but they may also cause inconsistencies in the data which render the feasible set empty.

Both these problems can be overcome if we consider a penalized version of the MaxEnt problem

$$\sup_{f \in L_1^+(D)} - \int_D f(x) \log f(x) dx - \frac{1}{2\theta} \sum_{i=1}^m w_i^2 \left[\int_D a_i(x, \bar{a}_i) f(x) dx - \tilde{a}_i \right]^2 \quad (2.6)$$

for some positive θ and w_i. Here θ plays the role of "temperature" — the lower we set θ, the smaller the errors will be. The w_i act as weights to emphasize the importance of a particular constraint — the higher we set w_i, the smaller the error

---

[5]The BFGS algorithm was independently developed by Broyden [3], Fletcher [8], Goldfarb [10] and Shanno [19].

[6]Borwein *et al.* [2] have however developed an easy-to-check test for this constraint qualification in the standard MaxEnt problem for call options.

for the ith constraint will be. Although this problem is unconstrained, as discussed in Decarreau *et al.* [7], we can reformulate (2.6) in the form of the original problem as

$$\sup_{f\in L_1^+(D),\epsilon\in\mathbb{R}^m} - \int_D f(x)\log f(x)dx - \frac{1}{2\theta}\|\epsilon\|^2$$

subject to

$$\int_D a_i(x,\bar{a}_i)f(x)dx - \tilde{a}_i = \frac{\epsilon_i}{w_i}, \quad i = 1,\dots,m,$$

where $\|\cdot\|$ denotes the Euclidean norm on $\mathbb{R}^m$. Then one can easily show that the dual function of this penalized problem is given by

$$\mathcal{L}^*(\lambda) = \log Z(\lambda) - \lambda'\tilde{a} + \frac{1}{2}\theta\|\tilde{\lambda}\|^2,$$

where $\tilde{\lambda} = (\lambda_1/w_1,\dots,\lambda_m/w_m)'$, and the ith element of its gradient vector is given by

$$\frac{\partial\mathcal{L}^*(\lambda)}{\partial\lambda_i} = \int_D a_i(x,\bar{a}_i)\hat{f}(x,\lambda)dx - \tilde{a}_i + \theta\frac{\lambda_i}{w_i^2}.$$

Thus we can still use the dual approach as discussed above to solve this problem.

The above technique is known as *penalization*. Another approach, called *relaxation*, is to solve the problem

$$\sup_{f\in L_1(D)} - \int_D f(x)\log f(x)dx$$

subject to

$$\left|\int_D a_i(x,\bar{a}_i)f(x)dx - \tilde{a}_i\right| \leq \epsilon, \quad i = 1,\dots,m$$

for some $\epsilon > 0$. The problem with relaxation, as opposed to penalization, is that the issue of consistency remains — the feasible set may still be empty if ϵ is too small. We will use penalization for our application.

2.4. *Principle of minimum relative entropy*

Since the distribution with the greatest entropy is the uniform distribution,[7] when we apply the principle of maximum entropy we are effectively choosing the distribution that is "closest" to the uniform distribution while satisfying the data constraints. But we could, if we wish, choose a distribution other than the uniform distribution. To do this we use the concept of *relative entropy*.

---

[7]Among distributions with bounded support.

Relative entropy is a measure of "distance" of one probability distribution to another. For absolutely continuous probability distributions F and G it is defined by[8]

$$L(F|G) = \int_D f(x) \log \frac{f(x)}{g(x)} dx,$$

where f and g are the densities of F and G respectively. Minimizing relative entropy will be useful if we have a *prior belief* about what the empirical distribution might or should be.

The form of the optimum solution to the *minimum relative entropy* (MinRelEnt) problem is almost identical to that of the MaxEnt problem and is given by

$$\hat{f}(x, \lambda) = \frac{1}{Z(\lambda)} g(x) \exp \left\{ \sum_{i=1}^{m} \lambda_i a_i(x, \bar{a}_i) \right\}, \tag{2.7}$$

where $Z(\lambda) := \int g(x) \exp\{\sum_i \lambda_i a_i(x, \bar{a}_i)\} dx$. The optimal f can again be found by solving for the unique values λ_i that satisfy the constraints (2.2), possibly using the dual approach as discussed above.

We need to note the regularity conditions required on the prior g. To apply the Euler–Lagrange equation, we must (corresponding to footnote 3) make the assumption that there is a feasible f with finite relative entropy which is equivalent to g (i.e., their supports agree almost everywhere). The latter is difficult to check but we can ensure that an optimal solution has finite relative entropy by requiring $\int_D |\hat{f}(x) \log \hat{f}(x)/g(x)| dx < \infty$. The actual conditions on g therefore depend on the functions a_i and the domain D.

3. The MaxEnt Copula Problem

If we wish to apply the MaxEnt principle to a problem involving a copula, then in addition to the data constraints (2.2) above we must also require constraints on the marginals of f.

A *copula* is a joint distribution function on the unit hypercube $[0, 1]^n$ with marginals that are uniformly distributed. So in addition to the data constraints, we need the *marginal constraints*

$$\int_0^p \int_{[0,1]^{n-1}} f(x) dx_{\neg j} dx_j = p \quad \forall p \in [0, 1] \quad j = 1, \ldots, n. \tag{3.1}$$

We therefore have an infinite dimensional constraint space, unlike the finite dimensional case defined by (2.2).

---

[8]Relative entropy is always non-negative and is equal to zero if and only if $f \equiv g$.

3.1. *Discrete approximation*

One technique to deal with the infinite dimensional constraints is to take only a finite number $p_1, \ldots, p_N \in [0, 1]$ of discrete points and require that the above marginal constraints hold only for this set of points, but not for all $p \in [0, 1]$. Thus we would have instead of (3.1)

$$\int_{p_{k-1}}^{p_k} \int_{[0,1]^{n-1}} f(x)dx_{\neg j}dx_j = \Delta p_k \quad j = 1, \ldots, n, \quad k = 1, \ldots, N,$$

where $\Delta p_k := p_k - p_{k-1}$. This method may be satisfactory for many applications. For example in pricing CDO tranches, if we take each p_k to correspond to quarterly time steps or even annual time steps and the longest maturity of the tranches is 10 years, then we need only take $N = 40$ or 10, respectively, because there is no need to go beyond the longest maturity in the data set. Furthermore, if we assume a homogeneous portfolio where each firm has the same marginal default time distribution, then we know that the Lagrange multipliers for the p_k's for each dimension will be the same. There will thus only be an additional 10 to 40 more constraints since we do not need a separate set of Lagrange multipliers for each dimension.

3.2. *The MaxEnt copula problem*

There are two alternative ways of formulating the MaxEnt copula problem.

For a copula with density c the MaxEnt problem is to maximize its entropy $-\int_{[0,1]^n} c(u) \log c(u)du$, subject to the data constraints being satisfied and the marginals of c being uniformly distributed. By a simple change of variables, we can show that this problem is equivalent to minimizing

$$\int_D f(x) \log \frac{f(x)}{f_1(x_1)f_2(x_2)} dx,$$

where f is the joint density function with support D, and f_1 and f_2 are the marginal density functions of the relevant problem (assuming here just two dimensions for simplicity). Thus we can see that maximizing the entropy of a copula is equivalent to mimizing the entropy of the corresponding joint distribution to be as close as possible to the *independent* case, i.e., $f(x) = f_1(x_1)f_2(x_2)$.

The other way to formulate the MaxEnt copula problem is to start from the joint distribution. Thus we maximize $-\int_D f(x) \log f(x)dx$ subject to the data constraints being satisfied and f having marginals f_1 and f_2. Expressing this in terms of the copula, the problem becomes equivalent to maximizing

$$-\int_{[0,1]^2} c(u) \log \{c(u)f_1(x_1)f_2(x_2)\} du,$$

where $x_i = F_i^{-1}(u_i)$ for $i = 1, 2$ and the F_i are the marginal cdfs.

It would seem that both ways of formulating the problem are equally valid, although the former allows for an easier interpretation.

4. Application to CDO Tranche Pricing

We now apply the principle of minimum relative entropy to CDO tranche pricing. There are several things to consider in specifying this problem.

4.1. *The empirical copula problem for CDO pricing*

We choose to work with the copula of default times rather than with the joint distribution. Writing the CDO tranche pricing equation (1.4) in integral form we have

$$\int_{[0,\infty)^n} D(\tau, T)f(\tau)d\tau - \bar{s}\int_{[0,\infty)^n} P(\tau, T)f(\tau)d\tau = 0. \tag{4.1}$$

Rewriting (4.1) using Sklar's lemma $(f(\tau) = c(u)\Pi f_i(\tau_i))$ yields

$$\int_{[0,1]^n} D\left(F^{-1}(u), T\right) c(u)du - \bar{s}\int_{[0,1]^n} P\left(F^{-1}(u), T\right) c(u)du = 0, \tag{4.2}$$

where in terms of the marginal default times $(F^{-1}(u) := (F_1^{-1}(u_1), \ldots, F_n^{-1}(u_n)) = (\tau_1, \ldots, \tau_n) = \tau)$ and $c(u)$ is the density of the copula of default times (noting that there is cancellation of the Jacobian factor).

Next we must determine the form of the functions a_i in terms of the argument u of the copula density c. Referring back the equations for the default and premium payoffs (1.2), (1.3) and using (4.2) we obtain[9]

$$a_i(\mathbf{u}, \bar{s}_i) = \sum_{j=1}^{n} e^{-r\tau_j} 1_{\{\tau_j \leq T\}} \left(M(\boldsymbol{\tau}_j, \boldsymbol{\tau}) - M(\boldsymbol{\tau}_j^-, \boldsymbol{\tau})\right)$$

$$- \bar{s}_i \sum_{k=1}^{P} e^{-rt_k} (\beta - \alpha - M(t_k, \boldsymbol{\tau})), \tag{4.3}$$

where $\boldsymbol{\tau}_j = F_j^{-1}(\mathbf{u}_j)$. As for the observed data values $\tilde{s}_i$ (1.4) implies that they are always zero except for the equity tranche.[10]

We can now specify the *empirical copula CDO pricing problem*. For computational efficiency we assume a homogenous portfolio. To account for bid-ask spreads and also for inaccuracy from Monte Carlo integration (discussed below) we solve the penalized minimum relative entropy problem as outlined in Sec. 2.3. Thus the problem is

$$\sup_{c \in L^1[0,1]^n, \epsilon \in \mathbb{R}^{m+N}} - \int_{[0,1]^n} c(u) \log \frac{c(u)}{\pi(u)} du - \frac{1}{2\theta}|\epsilon|^2$$

---

[9]We have not mentioned accrued payments for notational simplicity, but they have been included in the numerical studies.

[10]The quote for the equity tranche on the iTraxx and CDX indices represents the upfront payment that must be made, where the running premium is set at $\bar{a}_i = 500$ bps.

subject to the data constraints given by

$$\int_{[0,1]^n} a_i(u, \bar{s}_i) c(u) du - \tilde{s}_i = \frac{\epsilon_i}{w_i} \quad i = 1, \ldots, m$$

and the marginal constraints given by

$$\int_{p_{k-1}}^{p_k} \int_{[0,1]^{n-1}} c(u) du_{\neg j} du_j - \Delta p_k = \frac{\epsilon_k}{w_k} \quad k = 1, \ldots, N,$$

for each $j = 1, \ldots, n$, where $0 = p_0 < \cdots < p_N < 1$ and $\Delta p_k := p_k - p_{k-1}$.[11] Here π represents some *prior* copula for the specified marginal default time distributions. This problem has a unique solution given by

$$\hat{c}(u) = \frac{1}{Z(\lambda)} \pi(u) \exp \left\{ \sum_{i=1}^m \lambda_i a_i(u, \bar{s}_i) + n \sum_{k=1}^N \lambda_k 1_{\{p_k \leq u < p_{k+1}\}} \right\}, \quad (4.4)$$

where $\lambda \in \mathbb{R}^{m+N}$ with normalizing constant

$$Z(\lambda) = \int_{[0,1]^n} \pi(u) \exp \left\{ \sum_{i=1}^m \lambda_i a_i(u, \bar{s}_i) + n \sum_{k=1}^N \lambda_k 1_{\{p_k \leq u < p_{k+1}\}} \right\} du.$$

The *dual function* $\mathcal{L}^*$ of this problem is given by

$$\mathcal{L}^*(\lambda) = \log Z(\lambda) - \sum_{i=0}^n \lambda_i \tilde{s}_i - n \sum_{k=1}^N \lambda_k \Delta p_k + \frac{1}{2} \theta |\tilde{\lambda}|^2,$$

where $\tilde{\lambda} = (\lambda_1/w_1, \ldots, \lambda_{m+N}/w_{m+N})$, and again is strictly convex with gradient

$$\frac{\partial \mathcal{L}^*(\lambda)}{\partial \lambda_i} = \int_{[0,1]^n} a_i(u, \bar{s}_i) \hat{c}(u) du - \tilde{s}_i + \theta \frac{\lambda_i}{w_i^2}$$

$$\frac{\partial \mathcal{L}^*(\lambda)}{\partial \lambda_k} = n \int_{p_{k-1}}^{p_k} \int_{[0,1]^{n-1}} \hat{c}(u_{\neg j}, u_j) du_{\neg j} - n \Delta p_k + \theta \frac{\lambda_k}{w_k^2}.$$

We must impose the regularity condition on the prior copula π. As stated in Sec. 2.4 we require $\int |\hat{c}(u) \log \hat{c}(u)/\pi(u)| du < \infty$. If we set $\bar{c} := \hat{c}/\pi$, then we need

$$\int |\pi(u) \bar{c}(u) \log \bar{c}(u)| du < \infty. \quad (4.5)$$

Notice from (4.3) the functions a_i have bounded range. As each $u_j \to 0$ we have $a_i \to 125(1 - R)$. This corresponds to extreme the case where every firm has defaulted instantaneously at time $t = 0$. In the opposite extreme as each $u_j \to 1$ we have $a_i \to -\bar{s}_i PL_{\max}$ which corresponds to the case when no firms default. As a_i's are decreasing functions they are bounded. It follows that the range of $\bar{c}$ is $[b, B]$

---

[11]Note that we do not need to partition all of $[0, 1]$, as mentioned previously in Sec. 3.1. We can take p_N corresponding to the longest maturity in the data set.

for some $b > 0$ and $B < \infty$. Therefore

$$\int_{[0,1]^n} |\pi(u)\bar{c}(u) \log \bar{c}(u)| du \le \int_{[0,1]^n} |\pi(u)| du \cdot \text{ess sup}_u |\bar{c}(u) \log \bar{c}(u)| < \infty,$$

so we can see that (4.5) is satisfied for *any* prior copula π.

We have assumed that the portfolio is homogeneous, so that each firm has the same marginal default time distribution. Our experiments and also those of Hull and White [13] show that the value of a CDO tranche is not particularly sensitive to whether the portfolio is homogenous or heterogenous — it is the *average* of the default probabilities of all firms that principally determines the value of a CDO tranche.

4.2. *Numerical issues*

An important issue to consider is how to use the resulting empirical copula to compute CDO tranche prices. A typical CDO application would be to calibrate the copula to the iTraxx or CDX tranche quotes. The iTraxx and CDX indices both refer to a portfolio of 125 names. Therefore to compute the fair premium $\bar{s}$ using equation (4.2) requires computing $n = 125$ dimensional integrals. This forces us to use Monte Carlo methods.

With just simple Monte Carlo integration computing CDO prices with the Min-RelEnt copula is very slow to converge. Markov Chain Monte Carlo methods have been tried and converge much faster in some cases but are not robust in general.[12] However, the presence of the factor $\pi(x)$ in (4.4) allows us to use *importance sampling* if the prior is chosen to be a copula from which we can easily simulate. This is effective in speeding up convergence.

We also used an additional importance sampling technique proposed by Joshi [16]. We choose the prior π to be the Gaussian copula and generate random variates from it using the well-known one factor method. We then shift the mean of the common factor by some amount μ to simulate more default times that will affect the payoffs of the tranches and multiply the resulting integrand by the likelihood ratio $\exp\{0.5\mu^2 - \mu X\}$, where X is the realization of the common factor.

4.3. *Calibration to simulated CDO prices*

We conducted two sets of tests. The first set of tests involved generating a set of simulated market CDO quotes from a given copula. After calibrating the minimum relative entropy copula to a subset of these quotes — which we call the training set — we priced the remaining out-of-sample tranches with it and compared them to the known true prices to see how well the minimum relative entropy copula can learn about the true underlying copula.

---

[12]Specifically, the Metropolis random walk algorithm [17].

The following assumptions were used.

Number of firms	125
Risk-free rate	0.05
Hazard rate for each firm	0.005
Recovery rate	0.4
Premium payments per year	4

The prior for the minimum entropy copula was chosen to be the Gaussian copula with correlation 0.4.

We simulate the market CDO tranche quotes from the *stochastic correlation copula*. This copula is simply the Gaussian copula with random correlation values and is a good candidate because it is one of the copulas that fits market prices relatively well, as well as being easy to simulate from [5]. We will use the discrete distribution $\rho = (0.066, 0.2, 0.8)$ with probabilities $p = (0.66, 0.1, 0.24)$ for the random correlation parameter.[13]

The training set contains tranches of different maturities and all maturities are calibrated simultaneously. The training set values and the calibration results are given in Table 1. Note that the equity (0,3) tranche premia are not in basis points but are expressed as a percentage of the nominal to be paid upfront. As we can see, the "calibration" is very good.

The next step is to see how well the empirical copula can "interpolate" across tranche threshold levels. The out-of-sample tranche pricing results are shown in

Table 1. Calibration to simulated data.

Thresholds (%)	Maturity (yrs)	True premium (bps)	MinRel premium (bps)	Absolute error (bps)
0–3	5	14.7	14.7	0.0
3–6	5	99.2	99.6	0.4
6–9	5	32.9	33.3	0.5
9–12	5	21.8	22.1	0.3
12–22	5	14.0	13.9	0.1
0–3	7	18.2	18.3	0.1
3–6	7	136.2	136.3	0.1
6–9	7	39.7	39.9	0.2
9–12	7	23.3	23.5	0.2
12–22	7	14.6	14.6	0.0
0–3	10	21.3	21.4	0.1
3–6	10	185.1	185.6	0.5
6–9	10	53.9	54.0	0.1
9–12	10	26.7	26.7	0.0
12–22	10	15.4	15.3	0.1
Total error				2.7

[13]These are the values used in Burtschell *et al.* [5]. There is nothing inherently special about these values other than that they produce reasonable prices.

Table 2. Prices of non-standard tranches using the empirical copula.

Thresholds (%)	Maturity (yrs)	True premium (bps)	MinRel premium (bps)	Absolute error (bps)
1.5–4.5	5	271.1	276.7	5.6
4.5–7.5	5	49.6	50.3	0.7
7.5–10.5	5	25.9	26.2	0.3
10.5–17.0	5	17.2	17.0	0.2
1.5–4.5	7	331.6	336.9	5.3
4.5–7.5	7	65.8	66.1	0.3
7.5–10.5	7	28.7	29.5	0.8
10.5–17.0	7	17.8	18.2	0.4
1.5–4.5	10	389.5	396.9	7.4
4.5–7.5	10	94.1	92.9	1.2
7.5–10.5	10	35.4	34.8	0.6
10.5–17.0	10	19.0	19.0	0.0
Total error				22.8

Table 2. Although it has some trouble pricing the (1.5, 4.5) tranche accurately, overall we can see that the empirical copula performs well out-of-sample. For comparison, we also price these non-standard tranches using the industry standard base correlation approach [1]. The results are shown in Table 3. We can see that although the base correlation method is relatively accurate for the senior mezzanine and super senior tranches, it performs much worse than the MinRelEnt copula for the (1.5, 4.5) tranche. This is likely because the loss distribution associated with the true copula has a high peak in this region (see the discussion below). Overall, the MinRelEnt copula performs significantly better than the base correlation approach.

We also tested to see how well the empirical copula can "interpolate"/ "extrapolate" across maturities. The results are shown in Table 4. Again we can see excellent

Table 3. Prices of non-standard tranches using base correlation.

Thresholds (%)	Maturity (yrs)	True premium (bps)	BaseCorr premium (bps)	Absolute error (bps)
1.5–4.5	5	271.1	255.8	15.3
4.5–7.5	5	49.6	51.1	1.5
7.5–10.5	5	25.9	23.6	2.3
10.5–17.0	5	17.2	18.2	1.0
1.5–4.5	7	331.6	307.2	24.4
4.5–7.5	7	65.8	70.4	4.6
7.5–10.5	7	28.7	28.1	0.6
10.5–17.0	7	17.8	19.1	1.3
1.5–4.5	10	389.5	363.6	25.9
4.5–7.5	10	94.1	99.3	4.8
7.5–10.5	10	35.4	37.2	1.8
10.5–17.0	10	19.0	20.5	1.5
Total error				85.0

Table 4. Prices of tranches with non-standard maturities using the empirical copula.

Thresholds (%)	Maturity (yrs)	True premium (bps)	MinRel premium (bps)	Absolute error (bps)
0–3	3	9.7	9.2	0.5
3–6	3	64.8	70.0	5.2
6–9	3	28.7	34.4	5.7
9–12	3	20.6	25.5	4.9
12–22	3	13.2	18.5	5.3
0–3	4	12.4	12.2	0.2
3–6	4	81.2	82.0	0.8
6–9	4	30.6	32.2	1.6
9–12	4	21.3	22.4	1.1
12–22	4	13.7	15.1	1.4
0–3	6	16.6	16.6	0.0
3–6	6	117.5	117.1	0.4
6–9	6	35.8	35.8	0.0
9–12	6	22.3	22.0	0.3
12–22	6	14.1	14.1	0.0
0–3	8	19.5	19.4	0.1
3–6	8	153.5	151.8	1.7
6–9	8	43.8	43.6	0.2
9–12	8	24.1	24.7	0.6
12–22	8	14.8	14.6	0.2
0–3	9	20.5	20.5	0.0
3–6	9	169.8	170.0	0.2
6–9	9	48.4	49.0	0.6
9–12	9	25.2	26.1	0.9
12–22	9	15.0	15.2	0.2

performance of the empirical copula in interpolating across non-standard maturities. It even performs reasonably well extrapolating to maturities less than five years, although we can see the accuracy does start to deteriorate for the three-year maturity.

4.4. *Calibration to market CDO prices*

In the second set of tests, we examine whether or not the MinRelEnt copula can be accurately calibrated to market data. We first attempt to calibrate to each maturity separately and then to all maturities simultaneously to see how much accuracy is lost. If the calibration across all maturities is much poorer than the calibration for a single maturity, then this would suggest that the market does not price the dependency as static. We also calibrate to the 5 and 10 year maturities, and see how well these calibrations can match the 7 year prices out-of-sample.

The recovery rate is assumed to be the same as before, but the default rate curve is now determined from the market index quotes and is assumed to be piecewise constant between maturities. We also replace the constant risk-free rate used in the simulation experiment by the market swap curve.

Table 5. Calibration to iTraxx, each maturity separately.

Thresholds (%)	Maturity (yrs)	Bid-ask (bps)	Market premium (bps)	MinRel premium (bps)	Absolute error (bps)
0–3	5	0.5	23.3	24.5	1.2
3–6	5	2.0	68.0	69.6	1.6
6–9	5	2.0	19.0	18.8	0.2
9–12	5	3.0	9.5	8.9	0.6
12–22	5	0.5	4.3	4.1	0.2
0–3	7	0.5	43.1	43.9	0.8
3–6	7	4.0	202.0	203.4	1.4
6–9	7	3.0	46.5	45.5	1.0
9–12	7	2.0	24.0	22.9	1.1
12–22	7	2.0	9.0	7.9	1.1
0–3	10	0.5	54.3	54.3	0.0
3–6	10	10.0	580.0	580.4	0.4
6–9	10	6.0	117.0	117.1	0.1
9–12	10	3.0	51.5	51.6	0.1
12–22	10	2.0	20.0	20.1	0.1

Table 5 shows the calibration results for the iTraxx tranche quotes for 4 April 2006.[14] Although not quite as good as the results for the simulated data, we can see that the calibration errors lie mainly within the bid-ask spreads. Interestingly, it is again for the junior mezzanine (3,6) tranche where the MinRelEnt copula is least accurate. Similar performance was achieved for CDX data.

Next we attempted to calibrate to all maturities simultaneously, the results of which are shown in Table 6. The accuracy is noticeably worse than calibration to

Table 6. Calibration to iTraxx, all maturities simultaneously.

Thresholds (%)	Maturity (yrs)	Market premium (bps)	MinRel premium (bps)	Absolute error (bps)
0–3	5	23.3	21.0	2.3
3–6	5	68.0	71.9	3.9
6–9	5	19.0	25.7	6.7
9–12	5	9.5	14.5	5.0
12–22	5	4.3	7.7	3.4
0–3	7	43.1	43.7	0.6
3–6	7	202.0	208.1	6.1
6–9	7	46.5	53.6	7.1
9–12	7	24.0	27.5	3.5
12–22	7	9.0	12.2	3.2
0–3	10	54.3	56.0	1.7
3–6	10	580.0	590.7	10.3
6–9	10	117.0	127.7	10.7
9–12	10	51.5	54.9	3.4
12–22	10	20.0	22.8	2.8

[14] Data was kindly supplied by Credit Suisse.

single maturity data, which suggests that each maturity has a different underlying copula associated with the market tranche quotes. If this is the case then the fitted MinRelEnt copula is in a sense an "average" of the different underlying copulas.

Notwithstanding the poor fit, we also calibrated the MinRelEnt copula to the 5 and 10 year quotes only and priced 7 year tranches out-of-sample to see how well it matched market prices. The results are shown in Table 7. Apart from the super senior tranche — whose predicted premium is double the true value — we can see that the MinRelEnt copula has the ability to interpolate across maturities to within reasonable accuracy, even though overall the calibration results were poor.

In order to investigate further the issue of stationarity of the copula, we generated loss distributions from each of the MinRelEnt copulas calibrated to single maturity data and compared them visually. The *loss distribution* is just the probability distribution of the number of defaults in the portfolio for for a fixed time horizon.[15]

Figure 2 shows the 10 year loss distribution generated from the 10 year MinRelEnt copula. We include the loss distribution generated from the prior Gaussian copula for comparison.

We also generate the 5 and 7 year loss distributions from the 10 year MinRelEnt copula, and compare them to the 5 year loss distribution generated from the 5 year MinRelEnt copula and to the 7 year loss distribution generated from the 7 year MinRelEnt copula. These are shown in Figs. 3 and 4.

Table 7. Calibration to 5 and 10 year, and pricing 7 year.

Thresholds (%)	Maturity (yrs)	Market premium (bps)	MinRel premium (bps)	Absolute error (bps)
0–3	5	23.3	21.6	2.7
3–6	5	68.0	71.8	3.8
6–9	5	19.0	24.3	5.3
9–12	5	9.5	13.0	3.5
12–22	5	4.3	6.7	2.4
0–3	7	43.1	**42.8**	**0.3**
3–6	7	202.0	**216.8**	**14.8**
6–9	7	46.5	**53.2**	**6.7**
9–12	7	24.0	**25.8**	**1.8**
12–22	7	9.0	**18.1**	**9.1**
0–3	10	54.3	55.3	1.0
3–6	10	580.0	597.9	17.9
6–9	10	117.0	132.5	15.4
9–12	10	51.5	52.3	0.8
12–22	10	20.0	19.2	0.8

[15]That is to say it is the plot of the density function for the random variable given by equation (1.1) for a fixed time horizon $t = T$ and $R = 0$.

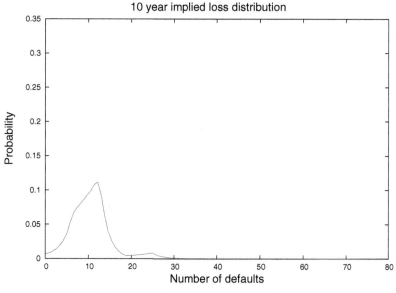

(a) Implied from 10 year MinRelEnt copula

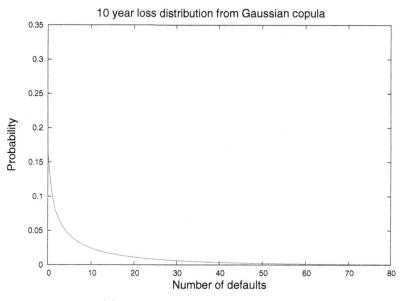

(b) Generated from prior Gaussian copula

Fig. 2. Loss distributions for the 10 year horizon.

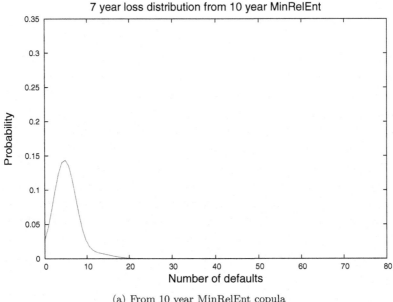

(a) From 10 year MinRelEnt copula

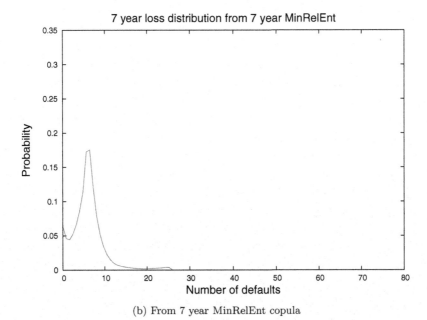

(b) From 7 year MinRelEnt copula

Fig. 3. Implied loss distributions for the 7 year horizon.

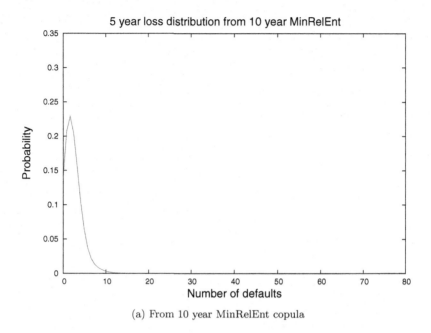

(a) From 10 year MinRelEnt copula

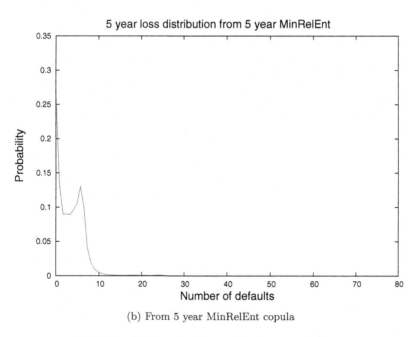

(b) From 5 year MinRelEnt copula

Fig. 4. Implied loss distributions for the 5 year horizon.

We can see some difference between the 7 loss distribution implied from the 10 year MinRelEnt copula and the 7 year loss distribution implied from the 7 year MinRelEnt copula. The difference is more pronounced in the implied 5 year loss distributions. This suggests that the market view on default dependence changes according to maturity, but in what way over time requires further research.

5. Conclusion

In this paper we have introduced a method to determine the minimum relative entropy copula and applied it to the pricing of CDO tranches. We calibrated the copula to market data, first by using tranches of only one maturity, and then to tranches across different maturities. Although we achieved excellent fits to single maturity data, the fit was noticeably worse for calibration across all maturites. However, when the same exercise was repeated using "market data" simulated from a known copula, we achieved a near-perfect fit across all maturities. Furthermore, we generated loss distributions from the empirical copulas implied from single maturity data and found that, for the same fixed time horizon, they were quite different from each other. These two observations suggest that the market view on default dependence may not be stationary across time.

The advantages of the entropic method are that it provides some justification for the choice of the copula, provides excellent fits to data and performs well out-of-sample. The entropic approach also allows us to empirically investigate whether or not default time dependency remains stationary across time.

There are two main disadvantages however. One is that like most copula methods it is assumed that when used for pricing, the dependence structure between default times remains static over time. As we have seen this may not be the case in reality. The other disadvantage is that both calibration and pricing involves computationally intensive procedures.

The entropic copula method is of course not limited in application to CDO tranche pricing but can be used wherever dependence is involved, e.g. for basket options.

References

[1] Bear Stearns, Valuing and hedging synthetic CDO tranches using base correlations, Technical Report (2004).

[2] J. M. Borwein, R. Choski and P. Maréchal, Probability distributions of assets inferred from option prices via the principle of maximum entropy, *SIAM Journal on Optimization* **14**(2) (2003) 464–478.

[3] C. G. Broyden, The convergence of a class of double rank minimization algorithms, *Journal of the Institute of Mathematics and its Applications* **6** (1970) 76–90 and 222–231.

[4] P. W. Buchen and M. Kelly, The maximum entropy distribution of an asset inferred from option prices, *Journal of Financial and Quantitative Analysis* **31**(1) (1996) 143–159.

[5] X. Burtschell, J. Gregory and J. P. Laurent, A comparative anaylsis of CDO pricing models, Working Paper, BNP Paribas (2005).

[6] S. Das, D. Duffie, N. Kapadia and L. Saita, Common failings: How corporate defaults are correlated, *Journal of Finance* **62**(1) (2007) 55–92.

[7] A. Decarreau, D. Hilhorst, C. Lemaréchal and J. Navaza, Dual methods in entropy maximization: Application to some problems in crystallography, *SIAM Journal on Optimization* **2**(2) (1992) 173–197.

[8] R. Fletcher, A new approach to variable metric algorithms, *Computer Journal* **13** (1970) 317–322.

[9] B. R. Frieden, *Physics from Fisher Information* (Cambridge University Press, Cambridge, 1998).

[10] D. Goldfarb, A family of variable metric methods derived by variational means, *Mathematics of Computation* **24** (1970) 23–26.

[11] G. D. Graziano and L. C. G. Rogers, A dynamic approach to the modelling of correlation credit derivatives using Markov chains, Working Paper, University of Cambridge (2006).

[12] R. J. Hawkins and B. R. Frieden, Financial probabilities from Fisher information, eprint, arXiv:cond-mat/0302579 (2006).

[13] J. Hull and A. White, Valuing credit derivatives using an implied copula approach, *Journal of Derivatives* **14**(2) (2006) 8–28.

[14] J. C. Jackwerth and M. Rubinstein, Recovering probability distributions from option prices, *Journal of Finance* **51** (1996) 1611–31.

[15] E. T. Jaynes, Information theory and statistical mechanics, *Physical Review* **106** (1957) 620–630.

[16] M. S. Joshi, Applying importance sampling to pricing single tranches of CDOs in a one-factor Li model, *Wilmott* **March** (2005).

[17] N. Metropolis, A. W. Rosenbluth, M. N. Rosenbluth and A. H. Teller, Equations of state calculations by fast computing machines, *Journal of Chemical Physics* **21**(1) (1953) 1087–1092.

[18] A. Mortensen, Semi-analytical valuation of basket credit derivatives in intensity-based models, *Journal of Derivatives* **13**(4) (2006) 8–26.

[19] D. F. Shanno, Conditioning of quasi-Newton methods for function minimization, *Mathematics of Computation* **24** (1970) 647–657.

PRICING AND HEDGING IN A DYNAMIC CREDIT MODEL

YOUSSEF ELOUERKHAOUI

CITIGROUP, Canada Square, Canary Wharf
London E14 5LB, United Kingdom
youssef.elouerkhaoui@citigroup.com

In this paper, we present a methodology for pricing and hedging portfolio credit derivatives in a dynamic credit model. Starting with a single-name Marshall–Olkin framework, we build a dynamic top-down version of the model, which is tractable and preserves the intuition of the original setting. In the first part of the paper, we derive analytically the Fourier transform of the loss variable and we study the skew dynamics implied by the model. In the second part, we develop a theory for dynamic hedging of portfolio credit derivatives. Since the market is incomplete, due to the residual correlation risk, perfect replication cannot be achieved. To find the hedging strategies, we use a quadratic risk minimization criterion.

Keywords: Marshall–Olkin model; common poisson shocks; dynamic copula; asymptotic series expansion; top-down approach; forward skew; marked point process; market incompleteness; dynamic hedging; quadratic risk minimization; Föllmer–Sondermann approach.

1. Introduction

Structured credit markets continue to expand and the demand for leveraged credit exposure is stronger than ever. Over the last couple of years, we have witnessed significant growth in the issuance of bespoke transactions and the liquidity in standard index tranches has improved dramatically. Index tranches are used by dealers to manage the correlation risk on their books and they offer various relative value opportunities for hedge funds and proprietary trading desks. The trend has been one of increased transparency. On the one hand, we have seen the beginnings of a market for tranchelets, which can be used to infer the interpolation (and extrapolation) assumptions of base correlation. On the other hand, the term structure dimension has emerged as an important feature that needs to be taken into account; today there is more issuance in seven-year and ten-year buckets, and most dealers show daily runs for three-year, five-year, seven-year and ten-year tenors.

Product innovation is also becoming one of the key drivers in the structured credit space. New products are customized to investors' needs and are designed to provide innovative ways of placing risk from different parts of the capital structure. Zero-Coupon Equity, Forward-Starting tranches, Leveraged Super Senior deals are some of the most recent examples. They are popular with investors for various reasons. For instance, there has been a strong interest from leveraged accounts for ten-year zero-coupon equity given the current low default environment and the tight premiums for equity risk; forward-starting tranches are used by investors to take views on the curve steepness and the actual realization of defaults; and the leveraged super senior deals, which are usually placed through conduits, have low mark-to-market volatility and until very recently had a very favorable rating treatment.

These recent developments in structured credit markets have motivated the need for more sophisticated models beyond the standard Gaussian copula framework. The key requirements for so-called third generation models are: (a) they should account for the term structure dimension and calibrate to all liquid tenors, (b) generate a reasonable "forward" skew behavior, (c) ensure arbitrage-free dynamics of the loss variable. In addition, to be of any practical use, the models need to have a tractable implementation and reasonable execution run times.

Above and beyond the fundamental flaw of the traditional copula approach, namely the fact that it is essentially a static model, it does not have well-defined replicating strategies *a la* Black-Scholes. As we evolve towards more mature dynamic models, hedging is going to be one of the key issues to be addressed.

A very promising modeling framework that was adopted by many researchers is the "Top-Down" approach. The recent literature on the topic includes: Bennani [2], Brigo *et al.* [4], Errais *et al.* [5], Giesecke and Goldberg [12], Schönbucher [18], Sidenius *et al.* [19]. Fundamentally, it is based on the idea that the loss variable is the key object that one needs to focus on. Instead of starting with the single-name information and construct the portfolio distribution, we would model directly the aggregate loss variable. As explained in Giesecke and Goldberg [12], losses through time can be viewed as a Marked Point Process whose distributional properties are fully determined by its compensator. By choosing different specifications of the loss compensator, we can, in principle, build a new family of dynamic loss models. This goes from a simple Poisson process or Compound Poisson to a richer time-changed (self-exciting) Hawkes process. In order to zoom-in on the single-name information, we would have to use the "random-thinning" technique.

This approach is conceptually very appealing since, by focusing on the aggregate level information, it offers a more natural calibration paradigm. However, for practical applications, it does suffer from a number of limitations. Firstly, it is not immediately obvious how we should define the loss compensator to match the market tranche quotes. In general, the calibration to multiple tenors can be quite challenging. Secondly, while random-thinning is a very powerful theoretical tool, it is not easy to implement. In particular, there is no simple way to get non-stochastic or mildly stochastic ratios that fit the single-name credit curves.

Our objective in this paper is twofold.

- First, build a dynamic credit model with the following features:
  - — Arbitrage-free loss density at every time horizon,
  - — Calibrates to all correlation skew tenors,
  - — Fast analytical implementation,
  - — Reasonable skew dynamics.

- Second, develop a self-consistent theory of dynamic hedging for exotic portfolio credit derivatives.

Our approach is to begin with a bottom-up setting, i.e., a suitable choice of single-name credit model, and overlay some dynamic properties on this static single-name framework. A suitable choice, here, would be one that has most of the features listed previously. Then, turning our attention to the aggregate portfolio information, we can infer the dynamics of the loss variable and make some reasonable simplifying assumptions to get a tractable SDE. We arrive thus at a new top-down model, which is both tractable and preserves the key properties of the original model and the intuition behind it.

Our starting point is the Marshall–Olkin model. Traditionally used in reliability theory (see [1]) to model the failure rate of multi-component systems, it was first introduced for credit derivatives pricing by Duffie [8]. Then, different variations of the model were analyzed by Duffie and Singleton [9], Duffie and Pan [8] and Duffie and Garleanu [7]. Its main underlying assumption is that the failure of each component or the default of each firm is triggered by a set of common independent Poisson processes. Note that this is one of the few models in the credit literature that allows for multiple instantaneous defaults. Incidentally, this is the reason why loss distributions, in this model, are generally "multi-modal", which facilitates the calibration to the tranche market. Furthermore, it is also known as a multivariate Poisson process and it has a number of useful analytical results that can be found, for instance, in Lindskog and McNeil [14].

The rest of the paper is structured as follows. In Sec. 2, we describe a dynamic version of the Marshall–Olkin model. In Sec. 3, we derive an asymptotic series expansion of the portfolio loss Fourier transform. In Sec. 4, we construct a top-down version of the model and we analyze the implied forward skew. In Sec. 5, we develop, in this framework, dynamic hedging strategies based on a quadratic risk minimization criterion.

2. The Model

2.1. *Model set-up*

We work on a probability space $(\Omega, \mathcal{G}, P)$, on which is given an $\mathbb{R}^d$-valued Itô process $X \triangleq (X_t)_{t \geq 0}$, describing the evolution of the state-variables in the economy. We denote by $\{\mathcal{F}_t\}$ the filtration generated by X: $\mathcal{F}_t \triangleq \sigma(X_s : 0 \leq s \leq t)$. Let

$(\tau_1, \ldots, \tau_n)$ be the default times of the obligors in the economy. We introduce, for each obligor i, the default indicator $D_t^i = 1_{\{\tau_i \leq t\}}$, which is equal to 1 if default occurs before time t and 0 otherwise. We denote by $\{\mathcal{H}_t^i\}$ the filtration generated by this process: $\mathcal{H}_t^i \triangleq \mathcal{F}_t^{D^i} = \sigma(D_s^i : 0 \leq s \leq t)$. The agents' filtration is the one generated by the economic state variables and the default processes

$$\mathcal{G}_t \triangleq \mathcal{F}_t \vee \left[\bigvee_{i=1}^n \mathcal{H}_t^i \right].$$

We assume that there exists a set of m independent Cox processes $(N^{c_j})_{1 \leq j \leq m}$, with independent intensities $(\lambda^{c_j}(X_t))_{1 \leq j \leq m}$, which can trigger one or multiple defaults. $(N^{c_j})_{1 \leq j \leq m}$ can be thought of as some systemic factors common to all names, or idiosyncratic events specific to each firm. Each process N^{c_j} can also be equivalently represented by the jump times $\{\theta_r^{c_j}\}_{r \in \{1,2,\ldots\}}$.

For each event of type c_j, and for all $t \geq 0$, we define a vector of independent Bernoulli variables $\mathbf{A}_t^j = (A_t^{1,j}, \ldots, A_t^{n,j})$ with probabilities $(p^{1,j}, \ldots, p^{n,j})$. We assume that, for $j \neq k$, the vectors $\mathbf{A}_t^j$ and $\mathbf{A}_t^k$ are independent. And we assume that, for $t \neq s$, the vectors $\mathbf{A}_t^j$ and $\mathbf{A}_s^j$ are independent. In other words, conditional on each trigger event of type c_j, at time $\theta_r^{c_j}$, we simulate a new vector of independent Bernoulli variables $(A_{\theta_r^{c_j}}^{1,j}, \ldots, A_{\theta_r^{c_j}}^{n,j})$. The variable $A_{\theta_r^{c_j}}^{i,j}$ indicates if issuer i has defaulted or not. In fact, we do not really need to define $(\mathbf{A}_t^j)$ for all $t \geq 0$, since the only times where the Bernoulli variables would need to be simulated are the common trigger events' jump times $\theta_r^{c_j}$. However, this was introduced for notational clarity when we formulate the SDE, which drives the single-name default dynamics.

We define the single-name processes $(N^i)_{1 \leq i \leq n}$ as

$$N_t^i \triangleq \sum_{j=1}^m \sum_{\theta_r^{c_j} \leq t} A_{\theta_r^{c_j}}^{i,j}, \quad \text{for } 1 \leq i \leq n. \tag{2.1}$$

The process N^i is also a Cox process since it is obtained as the sum of independent thinned Cox processes. Its intensity is given by $\lambda^i(X_t) = \sum_{j=1}^m p^{i,j} \lambda^{c_j}(X_t)$.

Now, we can define the single-name default time τ_i as the first jump time of the Cox process N^i:

$$\tau_i \triangleq \inf\{t : N_t^i > 0\}.$$

This completes the description the multivariate dependence of $(\tau_1, \ldots, \tau_n)$.

An alternative way of describing the model is by specifying the actual dynamics of the default processes. This was the approach used in Duffie [6]. Each default is decomposed on a set of "basic" events, which are pooled back together via $\{0,1\}$-variables to construct the "derived" single-name default events. The default point process D^i is assumed to be of the following form:

$$dD_t^i = (1 - D_{t-}^i) \sum_{j=1}^m A_t^{i,j} dN_t^{c_j}; \tag{2.2}$$

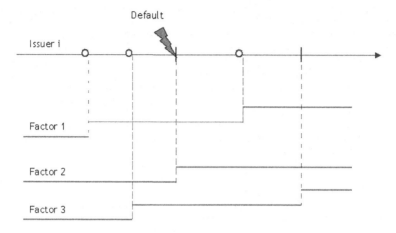

Fig. 1. Structure of the Marshall–Olkin model.

this stochastic differential equation should be understood as

$$D_t^i = \int_0^t (1 - D_{s-}^i) \sum_{j=1}^m A_s^{i,j} dN_s^{c_j}.$$

Throughout the paper, we use the convention that the first $m_c \triangleq m - n$ are common to all names, i.e., can trigger joint defaults, and the remaining n factors are specific to each firm. We denote the idiosyncratic factors by $(N^{0,i})_{1 \le i \le n}$ and Eq. (2.2) becomes

$$dD_t^i = \left(1 - D_{t-}^i\right) \left[\sum_{j=1}^{m_c} A_t^{i,j} dN_t^{c_j} + dN_t^{0,i} \right]. \tag{2.3}$$

Note that the Marshall–Olkin filtration is much larger than the one accessible to the agents in the economy. It contains the information flow of the common trigger events and the "conditional" Bernoulli events:

$$\widetilde{\mathcal{G}}_t \triangleq \mathcal{F}_t \vee \left[\bigvee_{j=1}^m \mathcal{F}_t^{N^{c_j}} \right] \vee \left[\bigvee_{j=1}^m \bigvee_{i=1}^n \mathcal{F}_t^{A^{i,j}} \right].$$

This is actually the reason why multi-name credit markets are, in general, incomplete. We have $\{\mathcal{G}_t\} \subset \{\widetilde{\mathcal{G}}_t\}$; the additional information not captured in the $\{\mathcal{G}_t\}$-filtration is the one associated to the multivariate dependence or the correlation risk. Additional market instruments are required to hedge this type of risk.

2.2. The equivalent fatal shock representation

In this subsection, we present another way of describing the Marshall–Olkin model: the "Equivalent Fatal Shock Representation". The common Poisson shocks $(N^{c_j})_{1 \le j \le m}$, introduced previously, have a clear and intuitive financial meaning.

They are, however, "non-fatal" in the sense that each trigger event may or may not lead to a single-name default. This can make the calculations of joint probability distributions fairly complicated, as one needs to keep track of the whole history of defaults after each trigger event. The Fatal Shock Model, on the other hand, introduces a new basis of common Poisson shocks, which facilitates these calculations. Its basic idea is the following. Instead of considering all types of shocks, we account only for the ones that induce defaults, hence, the name Fatal-Shock model. This new representation offers a very powerful tool for deriving some non-trivial results that will be used throughout.

Let $\mathbf{\Pi_n}$ be the set of all subsets of $\{1, \ldots, n\}$. For each subset $\pi \in \mathbf{\Pi_n}$, we introduce the point process N^π, which counts the total number of shocks in $(0, t]$ resulting in joint defaults of the obligors in π only:

$$N_t^\pi \triangleq \sum_{j=1}^{m} \sum_{r=1}^{N_t^{c_j}} I_{\theta_r^{c_j}}^{\pi,j}, \tag{2.4}$$

where, for each $t \geq 0$, the Bernoulli variable $I_t^{\pi,j}$ is defined as

$$I_t^{\pi,j} \triangleq \prod_{i \in \pi} A_t^{i,j} \prod_{i \notin \pi} (1 - A_t^{i,j}).$$

The following is the key result of the fatal shock representation.

Proposition 2.1. (*Fatal Shock Representation*). *For each subset $\pi \in \mathbf{\Pi_n}$, N^π is a Cox process with intensity*

$$\lambda^\pi(X_t) = \sum_{j=1}^{m} p^{\pi,j} \lambda^{c_j}(X_t), \quad \text{where } p^{\pi,j} \triangleq \prod_{i \in \pi} p^{i,j} \prod_{i \notin \pi} (1 - p^{i,j}).$$

Conditional on $\mathcal{F}_\infty$, the processes $(N^\pi)_{\pi \in \mathbf{\Pi_n}}$ are independent.

In the static case, the proof that the processes $(N^\pi)_{\pi \in \mathbf{\Pi_n}}$ are independent is nontrivial; it can be found, for example, in Lindskog and McNeil [14]. The extension to the dynamic version of the model is an immediate corollary.

The individual counting processes can, then, be expressed as

$$N_t^i = \sum_{\pi \in \mathbf{\Pi_n}} \mathbf{1}_{\{i \in \pi\}} N_t^\pi. \tag{2.5}$$

This is similar to the original model since the single-name processes are decomposed on a set of common independent Cox processes; the difference, however, is that each trigger event is fatal. We can see this by defining the equivalent Bernoulli variable $A_t^{i,\pi}$, which corresponds to a trigger event of name i conditional on an event of type π, as $A_t^{i,\pi} \triangleq \mathbf{1}_{\{i \in \pi\}}$,

$$N_t^i = \sum_{\pi \in \mathbf{\Pi_n}} \sum_{r=1}^{N_t^\pi} A_{\theta_r}^{i,\pi} = \sum_{\pi \in \mathbf{\Pi_n}} \mathbf{1}_{\{i \in \pi\}} N_t^\pi.$$

This new representation will be used, for instance, to derive the copula function implied by the model and the Marked Point Process of Sec. 5.

2.3. Dynamic copula

The existing theory of copula functions, which goes back to Sklar [20], has been extended by Patton [16] to conditional laws. He has introduced the so-called "conditional copula" to study dynamic dependence structures. In particular, he has used a Gaussian conditional copula with a time-dependent parameter to analyze the dependence between currency exchange rates. Starting with conditional distributions with respect to a given filtration, the conditional copula can be used to map a set of conditional marginals to the corresponding conditional multivariate distribution, and hence specify the dynamics of the multivariate model.

In this subsection, we establish the time-dependent copula of the dynamic model described above, and we discuss briefly its parameterization and the application to the CDO tranche market.

First, we state the equivalent form of Sklar's theorem for conditional copulas and we refer to Patton [16] for additional details.

Theorem 2.1. *Let $\mathcal{A}$ be some arbitrary sub-σ-algebra. Let $H(|\mathcal{A})$ be an n-dimensional conditional distribution function with conditional margins $F_1(|\mathcal{A}), \ldots, F_n(|\mathcal{A})$. Then, there exists a conditional n-copula $C(|\mathcal{A}) : [0,1]^n \times \Omega \to [0,1]$ such that for almost every $\omega \in \Omega$, and for all $(x_1, \ldots, x_n) \in \overline{\mathbb{R}}^n$,*

$$H(x_1, \ldots, x_n | \mathcal{A})(\omega) = C(F_1(x_1 | \mathcal{A})(\omega), \ldots, F_n(x_n | \mathcal{A})(\omega) | \mathcal{A})(\omega).$$

If for almost every $\omega \in \Omega$, the functions $x_i \to F_i(x_i | \mathcal{A})(\omega)$ are all continuous, then $C(|\mathcal{A})(\omega)$ is unique.

A natural choice for the sub-σ-algebra is $\mathcal{A} = \mathcal{F}_t$. The conditional copula $C(|\mathcal{F}_t)(\omega)$ that links the default times conditional marginals $\mathbb{P}(\tau_i > T_i | \mathcal{F}_t)(\omega)$ to their n-variate conditional distribution $\mathbb{P}(\tau_1 > T_1, \ldots, \tau_n > T_n | \mathcal{F}_t)(\omega)$ is a dynamic "copula process", which offers an alternative description of the dynamic Marshall–Olkin model.

The latter can be derived from the equivalent fatal shock representation. Indeed, using the $\mathcal{F}_\infty$-conditional independence of the processes $(N^\pi)_{\pi \in \mathbf{\Pi_n}}$, and the law of iterated expectations, we obtain the following $\mathcal{F}_t$-conditional multivariate probability function.

Proposition 2.2. (*Conditional Multivariate Probability Function*).

$$\mathbb{P}(\tau_1 > T_1, \ldots, \tau_n > T_n | \mathcal{F}_t) = \mathbb{E}\left[\exp\left(- \sum_{\pi \in \mathbf{\Pi_n}} \Lambda^\pi_{\max_{i \in \pi}(T_i)} \right) | \mathcal{F}_t \right], \qquad (2.6)$$

where the process Λ^π is the cumulative intensity: $\Lambda^\pi_T \triangleq \int_0^T \lambda^\pi(X_s) ds$.

This is the dynamic version of the Multivariate Exponential Distribution developed by Marshall–Olkin (1967).

In a typical parameterization of the model, we usually consider four types of market factors: 1. Global (or "World"), 2. Beta, 3. Sector, 4. Idiosyncratic. The

Global factor or World driver represents the Super-Senior risk in the portfolio: it has a low-default probability, but it triggers the default of all issuers simultaneously. The Beta factor introduces a correlation between credits in different sectors; and the sector drivers are used to add more correlation for credits within the same industry bucket. The idiosyncratic factors represent the firm-specific risks and drive the value of equity tranches.

$$\lambda_t^i = \left[\lambda_t^W\right] + p^{i,B}\left[\lambda_t^B\right] + \sum_{j=1}^{m_c-2} p^{i,S_j}\left[\lambda_t^{S_j}\right] + \left[\lambda_t^{0,i}\right].$$

In general, the factor loadings are usually chosen so that the contribution of each factor is proportional to the total intensity. If the factor loading $p^{i,j}$ is at its maximal value of one, all the incremental changes will be due to the idiosyncratic factor. The levels of the market factors are calibrated to liquid market instruments such CDO tranches of iTraxx and CDX. The Marshall–Olkin portfolio loss distribution, depicted in Fig. 2, is multi-modal. The first mode corresponds to the idiosyncratic contributions; the second mode is due to the Beta factor; the third hump in the middle is due to the sector factors; and the last spike at the end is because of the World driver. This multi-modality offers enough flexibility for getting a reasonably good fit of the base correlation skew.

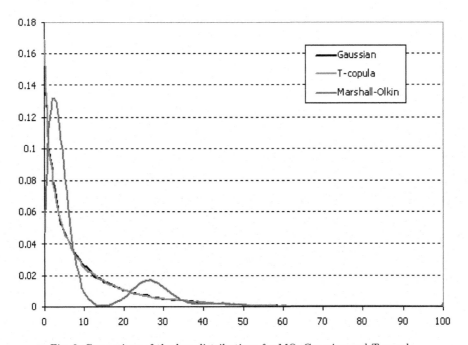

Fig. 2. Comparison of the loss distributions for MO, Gaussian and T-copula.

3. Numerical Implementation

The key ingredient in credit portfolio modeling is the distribution, at a given time horizon T, of the loss variable:

$$L_T \triangleq \sum_{i=1}^{n} L_i D_T^i,$$

where L_i is the loss upon default of issuer i. This can be computed in a number of ways:

- Monte-Carlo Method,
- Poisson Approximation,
- Duffie's Approximation,
- Asymptotic Series Expansion.

In this section, we focus on the analytical solutions. First, we review the Poisson approximation and Duffie's approximation, and we benchmark the numerical accuracy of each method. We shall see that for some parts of the capital structure, neither method is satisfactory. This will motivate the need for a better closed form solution. Our contribution here is to derive an exact asymptotic series expansion, which offers an accurate and fast analytical implementation.

3.1. *Poisson approximation*

In Lindskog and McNeil [14], the default indicators D_T^i are approximated by their corresponding Poisson counters N_T^i. This can be a reasonable approximation for low default probabilities:

$$L_T \simeq \sum_{i=1}^{n} L_i N_T^i \triangleq Z_T. \tag{3.1}$$

Replacing the N_T^i's by their Marshall–Olkin decomposition, we find that Z_T is a compound Poisson process, which is obtained as the sum of m independent compound Poisson processes $Z_T^{c_j}$:

$$Z_T^{c_j} \triangleq \sum_{r=1}^{N_T^{c_j}} \sum_{i=1}^{n} L_i A_{\theta_r^{c_j}}^{i,j} = \sum_{r=1}^{N_T^{c_j}} L_r^{c_j}. \tag{3.2}$$

Each $Z_T^{c_j}$ is defined by its Poisson counter $N_T^{c_j}$ and its compounding distribution $L_r^{c_j}$. When the jump sizes $L_r^{c_j}$ are discrete random variables taking values in some set $\{a_1, a_2, \ldots\}$, we can write its Laplace transform as

$$\mathcal{L}_{Z_T^{c_j}}(\alpha) = \mathbb{E}\left[e^{-\alpha Z_T^{c_j}}\right] = \mathbb{E}\left[\exp\left(-\Lambda_T^{c_j}\left[\sum_k \left(1 - e^{-\alpha a_k}\right)\mathbb{P}\left(L^{c_j} = a_k\right)\right]\right)\right]$$

$$= \mathcal{L}_{\Lambda_T^{c_j}}\left(\sum_k \left(1 - e^{-\alpha a_k}\right)\mathbb{P}\left(L^{c_j} = a_k\right)\right).$$

The distribution of the conditional losses $L^{c_j} = \sum_{i=1}^{n} L_i A^{i,j}$ can be computed by Fourier inversion or by using the recursion algorithm. The Laplace transform of the loss variable is then approximated by:

$$\mathcal{L}_{L_T}(\alpha) \simeq \prod_{j=1}^{m} \mathcal{L}_{\Lambda_T^{c_j}}\left(\sum_k \left(1 - e^{-\alpha a_k}\right) \mathbb{P}\left(L^{c_j} = a_k\right)\right). \tag{3.3}$$

The loss distribution is recovered by inverting the Laplace transform. This method can be used for all intensity processes whose Laplace transform is known analytically. Other methods, such as Panjer recursion, can also be used in the static case (see, for example, [14]) or with special specifications of the intensity process.

3.2. Duffie's approximation

Duffie and Pan neglect the probability of multiple jumps of the common market events and assume that the solution of the default SDE is given by

$$D_T^i \simeq \sum_{j=1}^{m} A^{i,j} D_T^{c_j}. \tag{3.4}$$

We can then approximate the Laplace transform of the loss variable as

$$\mathcal{L}_{L_T}(\alpha) \simeq \prod_{j=1}^{m} \left[\mathbb{E}\left[\exp\left(-\Lambda_T^{c_j}\right)\right] + \left(1 - \mathbb{E}\left[\exp\left(-\Lambda_T^{c_j}\right)\right]\right) \mathcal{L}_{L^{c_j}}(\alpha)\right], \tag{3.5}$$

where the Laplace transform of the conditional loss $L^{c_j} = \sum_{i=1}^{n} L_i A^{i,j}$ is given by:

$$\mathcal{L}_{L^{c_j}}(\alpha) \simeq \prod_{i=1}^{n} \left(p^{i,j} e^{-\alpha L_i} + \left(1 - p^{i,j}\right)\right).$$

In Fig. 3, we benchmark the numerical accuracy of the two approximations on a portfolio of 100 names. We price thin tranches across the capital structure, which correspond to a first-to-default, second-to-default,..., 100th-to-default. The errors are relative differences of break-even spreads.

Both methods are good approximations for the two extremes of the spectrum, but mis-price the middle part. As expected the Poisson approximation overestimates the value since by replacing the default indicator with a Poisson process we are adding more defaults. On the hand, Duffie's approximation under-estimates the value as we are approximating the common market factor Poisson counters with their corresponding default indicators, hence there are less common market factor triggers and less defaults.

3.3. Asymptotic series expansion

In this subsection, we derive an exact analytic formula of the Fourier transform of the loss variable. We shall use the following lemma.

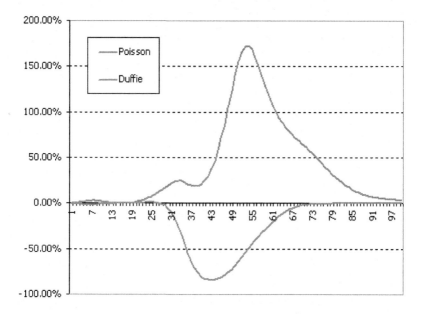

Fig. 3. Numerical accuracy of the Poisson approximation and Duffie's approximation.

Lemma 3.1. *The Fourier transform* $\phi(u) \triangleq \mathbb{E}[\exp(iuL_T)]$ *of the loss variable* L_T *is given by*

$$\phi(u) = \sum_{\pi \in \Pi_n} Q_\pi^{[1]}(T) \left[\prod_{i \notin \pi} \psi_i \prod_{i \in \pi} (1 - \psi_i) \right], \qquad (3.6)$$

where $\psi_i \triangleq \exp(iuL_i)$ *and the factor* $Q_\pi^{[1]}(T)$ *is the survival probability of all the names belonging to the subset* π.

We denote by

$$\phi(u, \widetilde{p}) \triangleq \prod_{i=1}^{n} ((1 - p(i)) + p(i) \exp(iuL_i)),$$

the Fourier transform of the variable $\sum_{i=1}^{n} L^i Y_i$, where $(Y_1, \ldots, Y_n)$ are n independent Bernoulli variables with probabilities $\widetilde{p} = (p(1), \ldots, p(n))$.

The factors $Q_\pi^{[1]}(T)$ can be computed in closed form as:

$$Q_\pi^{[1]}(T) = \mathbb{E} \left[\prod_{i \in \pi} (1 - D_T^i) \right] = \mathbb{E} \left[\exp \left(-\sum_{j=1}^{m} \Lambda_T^{c_j} \left(1 - \prod_{i=1}^{n} (1 - p^{i,j}) \right) \right) \right].$$

By expanding the $Q_\pi^{[1]}(T)$-factors, we obtain the following series expansion.

Theorem 3.1. (*Asymptotic Series Expansion*). *The loss Fourier transform $\phi(u)$ has the following series expansion*

$$\phi(u) = \sum_{n_1=0}^{+\infty} \cdots \sum_{n_{m_c}=0}^{+\infty} \mathbb{E}\left[e^{-\Lambda_T^{c_1}}\frac{(\Lambda_T^{c_1})^{n_1}}{n_1!}\right]$$

$$\cdots \mathbb{E}\left[e^{-\Lambda_T^{c_{m_c}}}\frac{(\Lambda_T^{c_{m_c}})^{n_{m_c}}}{n_{m_c}!}\right] \phi\left(u, \widetilde{p_{n_1,\ldots,n_{m_c}}}\right), \tag{3.7}$$

where the probability vector $\widetilde{p_{n_1,\ldots,n_{m_c}}}$ is given by

$$p_{n_1,\ldots,n_{m_c}}(i) \triangleq 1 - \mathbb{E}\left[e^{-\Lambda_T^{0,i}}\right]\left(1 - p^{i,1}\right)^{n_1} \cdots \left(1 - p^{i,m_c}\right)^{n_{m_c}}.$$

This can be seen as the formula of a "Binomial Mixture". The first term, i.e., for $n_1 = \cdots = n_{m_c} = 0$, corresponds to the idiosyncratic mode, that is when all the names are independent and have default probabilities $(1 - \mathbb{E}[e^{-\Lambda_T^{0,i}}])_{1 \leq i \leq n}$. The additional terms are overlaid on the independent case to build out the correlation profile. Each correlation mode corresponds to various market factor contributions. Note that when n is large, each Binomial distribution in the expansion converges to a Gaussian and Eq. (3.7) becomes the formula of a "Gaussian Mixture".

4. Dynamic Loss Distribution

In this section, we derive the dynamics of the loss variable L_t, defined as:

$$L_t \triangleq \frac{1}{n}\sum_{i=1}^{n}(1-\delta)D_t^i.$$

First, we construct a top-down version of the Marshall–Olkin model. Then, we apply the model to Forward-Starting tranches and we study the behavior of the forward skew.

4.1. *Dynamics of the loss variable*

Using the SDE of the individual default indicators, we get

$$dL_t = ((1-\delta) - L_{t-})\sum_{j=1}^{m}\left[\frac{\frac{1-\delta}{n}\sum_{i=1}^{n}\left(1 - D_{t-}^i\right)A_t^{i,j}}{(1-\delta) - L_{t-}}\right]dN_t^{c_j}. \tag{4.1}$$

We introduce the coefficient $\gamma_t^{c_j}(\omega)$, defined as

$$\gamma_t^{c_j}(\omega) \triangleq \frac{\frac{1-\delta}{n}\sum_{i=1}^{n}\left(1 - D_{t-}^i(\omega)\right)A_t^{i,j}(\omega)}{(1-\delta) - L_{t-}(\omega)}.$$

We have added the explicit dependence on ω, to emphasize the fact that this fraction is a stochastic object. Eq. (4.1) becomes

$$dL_t = ((1-\delta) - L_{t-})\sum_{j=1}^{m}\gamma_t^{c_j}(\omega)dN_t^{c_j}.$$

In the following, we shall approximate the stochastic fraction $\gamma_t^{c_j}(\omega)$ with a fixed deterministic amount. In other words, we assume that all the path-dependency is included in the loss term. We shall proceed in two steps.

First, we remove the path dependence by approximating $\gamma_t^{c_j}(\omega)$ with realizations where the default process and loss process are $D_t^i(\omega) = 0$ and $L_t(\omega) = 0$. This is similar to the idea used in Duffie's approximation to simplify the dynamics of the model:

$$\gamma_t^{c_j}(\omega) \triangleq \frac{\frac{1-\delta}{n}\sum_{i=1}^{n}\left(1 - D_{t-}^i(\omega)\right)A_t^{i,j}(\omega)}{(1-\delta) - L_{t-}(\omega)} \sim \frac{\sum_{i=1}^{n}A_t^{i,j}(\omega)}{n}.$$

Then, we neglect the variance of the approximated ratio $\gamma_t^{c_j}(\omega)$ and replace it by its mean,

$$\frac{\sum_{i=1}^{n}A_t^{i,j}(\omega)}{n} \simeq \frac{\sum_{i=1}^{n}p^{i,j}}{n} \triangleq \gamma^{c_j}.$$

We arrive at the following SDE

$$dL_t = ((1-\delta) - L_{t-})\left[\sum_{j=1}^{m_c}\gamma^{c_j}dN_t^{c_j} + \sum_{i=1}^{n}\gamma^{0,i}dN_t^{0,i}\right],$$

which describes a simpler and more tractable dynamic of the model. This can be simplified even further by aggregating all the idiosyncratic terms. Indeed, by observing that all the idiosyncratic fractions $\gamma^{0,i}$ are equal, $\gamma^{0,i} = \gamma^0 = \frac{1}{n}$, and that the process N^0, defined as $N_t^0 \triangleq \sum_{i=1}^{n}N_t^{0,i}$, is a Poisson process with intensity $\sum_{i=1}^{n}\lambda_t^{0,i}$, we can re-write the SDE as follows:

$$dL_t = ((1-\delta) - L_{t-})\left[\sum_{j=1}^{m_c}\gamma^{c_j}dN_t^{c_j} + \gamma^0 dN_t^0\right].$$

We can do a similar analysis for the sector drivers, since all the sector fractions have, roughly, the same order of magnitude:

$$\gamma^{Sector} = \frac{1}{n} \times \{\# \text{ credits in sector}\} \times \{\text{average } p^{i,S}\}.$$

To sum up. We have constructed a new top-down model, which is described by the dynamics of the loss variable and preserves the intuition of the original Marshall–Olkin framework:

$$dL_t = ((1-\delta) - L_{t-})\left[\sum_{j=1}^{m}\gamma^j dN_t^j\right]. \tag{4.2}$$

This is the SDE studied empirically in Longstaff and Rajan [15]. They have investigated how this description of the loss variable performs against historical data. Some of their findings include the fact that the CDO tranche market is best described with three factors, which can be interpreted as a firm specific, industry and economy-wide default events. They have also found that, on average, 65% of the index spread is due to firm-specific risk, 27% is sector risk and 8% is systemic risk.

Furthermore, it is interesting to note that the SDE of the Generalized Poisson Loss model proposed by Brigo *et al.* [4] is very close to our top-down Marshall–Olkin. However, since they decompose the loss variable on a set of Poisson processes directly, it is possible for L_t to exceed its maximal value. To circumvent the issue, they have added the min(.) operator to their representation. In our case, the SDE is well posed and L_t is bounded; the term $((1 - \delta) - L_{t-})$ in front of the Poisson shocks plays the role of an absorbing barrier, which pulls back the solution towards $(1 - \delta)$.

The solution of the SDE is obtained by applying Itô's lemma for jump processes (see [17]). Indeed, if we consider the intermediate process $Y_t \triangleq \log((1 - \delta) - L_t)$, then Itô's lemma gives

$$dY_t = \sum_{j=1}^{m} \log(1 - \gamma^j) \Delta N_t^j,$$

which leads to the solution of the SDE:

$$L_t = (1 - \delta) \left[1 - e^{\sum_{j=1}^{m} \log(1-\gamma^j) N_t^j} \right] \simeq (1 - \delta) \left[1 - e^{-\sum_{j=1}^{m} \gamma^j N_t^j} \right]. \qquad (4.3)$$

To lighten up the notations, we have approximated $\log(1 - \gamma^j) \simeq \gamma^j$ as the jump sizes are usually small in practice. The Fourier transform of the loss variable is given by

$$\phi(u) = \sum_{n_1=0}^{+\infty} \cdots \sum_{n_m=0}^{+\infty} \prod_{j=1}^{m} \mathbb{E} \left[e^{-\Lambda_T^j} \frac{\left(\Lambda_T^j \right)^{n_j}}{n_j!} \right] \exp \left(iu (1 - \delta) \left[1 - e^{-\sum_{j=1}^{m} \gamma^j n_j} \right] \right).$$

$$(4.4)$$

One of the advantages of this model is that it can be implemented for any choice of intensity dynamics as long as the Laplace transform of Λ_T^j is known analytically. A very popular choice, used for example in Duffie and Pan [8] and Longstaff and Rajan [15], is the CIR process:

$$d\lambda_t^j = \kappa^j \left(\lambda_\infty^j - \lambda_t^j \right) dt + \sigma^j \sqrt{\lambda_t^j} dW_t^j.$$

Its Laplace transform is given by

$$\mathcal{L}_{\Lambda_T^j}(x) = \exp \left(\alpha(T) + \beta(T) \lambda_0^j \right).$$

Alternatively, the Poisson probabilities can also be computed as

$$\mathbb{E} \left[e^{-\Lambda_T^j} \frac{\left(\Lambda_T^j \right)^{n_j}}{n_j!} \right] = A(T) \exp \left(-B(T) \lambda_0^j \right) \sum_{k=0}^{n_j} C_{n_j,k}(T) \left(\lambda_0^j \right)^k,$$

where the coefficients $C_{n_j,k}(T)$ are the solutions of a simple system of recursive ODEs that can be solved numerically.

4.2. *Application: Forward skew*

The payoff of a Forward-Starting CDO is defined as a call-spread on the forward loss variable $L_{T_1,T_2} \triangleq L_{T_2} - L_{T_1}$:

$$M_{T_1,T_2}^{(K_1,K_2)} = \min\left(\max\left(L_{T_1,T_2} - K_1, 0\right), K_2 - K_1\right).$$

One can evaluate this payoff by using a Black-type approach. Indeed, given the expected value of the forward loss

$$\mathbb{E}\left[L_{T_1,T_2}\right] = \mathbb{E}\left[L_{T_2}\right] - \mathbb{E}\left[L_{T_1}\right],$$

we can use a Gaussian copula and an exogeneous base correlation curve to define the distribution of L_{T_1,T_2}. In other words, this would specify how the expected loss $\mathbb{E}[L_{T_1,T_2}]$ is allocated across the capital structure. This base correlation curve will be referred to as the Forward Skew.

Definition 4.1. (Forward Skew). We call a forward skew the base correlation curve of the loss variable $L_{T_1,T_2} = L_{T_2} - L_{T_1}$.

This terminology is consistent with a standard practice in options market. To price a spread-option, a normal Black model is used where the spread, $S_{T_1,T_2} = S_{T_2} - S_{T_1}$, is assumed to follow a normal diffusion and the volatility parameter is referred to as the forward vol.

The top-down model, described by Eq. (4.2), is, naturally, suitable for forward tranches. Indeed, the forward loss variable is given by the expression

$$L_{T_2} - L_{T_1} = ((1 - \delta) - L_{T_1})\left[1 - \exp\left(-\sum_{j=1}^{m} \gamma^j \left(N_{T_2}^j - N_{T_1}^j\right)\right)\right], \qquad (4.5)$$

and in order to price a forward tranche, we simply need to evaluate two expectations. First, we compute the conditional expectation on L_{T_1}:

$$\mathbb{E}\left[f\left(L_{T_2} - L_{T_1}\right)|L_{T_1}\right] = \sum_{n_1=0}^{+\infty} \cdots \sum_{n_m=0}^{+\infty} \prod_{j=1}^{m} \mathbb{E}\left[e^{-\left(\Lambda_{T_2}^j - \Lambda_{T_1}^j\right)} \frac{\left(\Lambda_{T_2}^j - \Lambda_{T_1}^j\right)^{n_j}}{n_j!}\right]$$

$$\times f(((1-\delta) - L_{T_1})[1 - e^{-\sum_{j=1}^{m} \gamma^j n_j}]).$$

Then, we integrate with respect to the L_{T_1}-density

$$\mathbb{E}[f(L_{T_2} - L_{T_1})] = \sum_{n_1=0}^{+\infty} \cdots \sum_{n_m=0}^{+\infty} \prod_{j=1}^{m} \mathbb{E}\left[e^{-\Lambda_{T_1}^j} \frac{\left(\Lambda_{T_1}^j\right)^{n_j}}{n_j!}\right]$$

$$\times \mathbb{E}[f(L_{T_2} - L_{T_1})|L_{T_1} = (1 - \delta)[1 - e^{-\sum_{j=1}^{m} \gamma^j n_j}]].$$

In the following, we apply these formulas to analyze the behavior of forward tranches and the model implied forward skew.

To begin with, we consider a model with flat parameters: $\gamma^1 = 0.0048$, $\gamma^2 = 0.048$, $\gamma^3 = 0.6$, $\lambda^1 = 0.8$, $\lambda^2 = 0.02$, $\lambda^3 = 0.001$. We have the following break-even spreads for the 5 year, 10 year and 5 into 5 tranches.

	5y	5v5	10y
0-3	43.9	34.6	62.7
3-6	238.6	227.9	476.9
6-9	55.1	48.7	138.5
9-12	15.5	14.7	35
12-22	10	10	11.5

The main point to highlight here is the time-invariance property of the Marshall–Olkin model. For a Poisson process, it is a well-known fact that what happens between T_1 and T_2 conditional on the information at T_1 is similar to what happens between 0 and $(T_2 - T_1)$. This is known as the time-invariance or the memory-less property. The multivariate Poisson process of the Marshall–Olkin model is also memory-less. Hence, the break-even spreads of the spot 5 year tranches and the forward starting 5 into 5 are of the same order of magnitude. When we introduce a term structure, this is not true anymore.

To study the term structure effect, we calibrate the spot skews at two different tenors and we analyze the model implied forward skew.

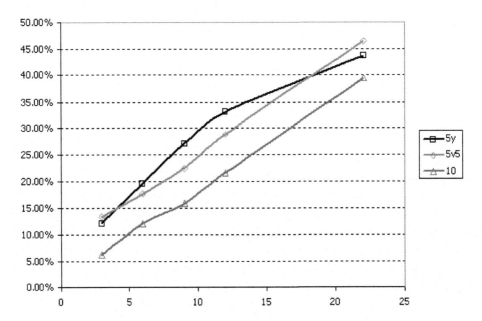

Fig. 4. Comparison of the spot 5 year and 10 year skews with 5 into 5 forward skew.

In Fig. 4, we match the 5 year and 10 year skews and compare them with the model implied 5 into 5 forward skew.

In Fig. 5, we match the 3 year and 10 year skews and compare them with the model implied 3 into 7 forward skew.

In Fig. 6, we match the 7 year and 10 year skews and compare them with the model implied 7 into 3 forward skew.

In the first example, the 5 into 5 forward skew is between the spot 5 year skew and the spot 10 year skew; the model offers a way of interpolating between the two skews. In the second example, the 3 into 7 forward skew is very close to the 10 year skew; this is not surprising since the expected loss for the shorter maturity is quite small, the two variables $(L_{10y} - L_{3y})$ and L_{10y} are not very dissimilar. In the limit case where T_1 goes to 0, we know that $(L_{T_2} - L_{T_1})$ converges to L_{T_2}. For practical applications, when the shorter maturity is small, one can use the skew curve for the longer maturity as a good proxy of the forward skew. In the third example, we have the exact opposite where the 7 year and 10 year skews are very similar since the two tenors are close to each other, and the 7 into 3 forward skew behaves more like the 3 year spot skew.

5. Dynamic Hedging

In this section, we develop a theory for dynamic hedging of portfolio credit derivatives in the Marshall–Olkin framework. To address the market incompleteness, we use a quadratic risk minimization criterion.

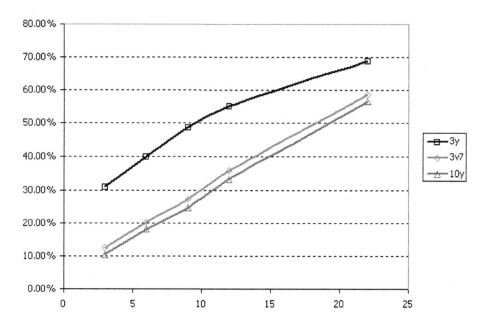

Fig. 5. Comparsion of the spot 3 year and 10 year skews with the 3 into 7 forward skew.

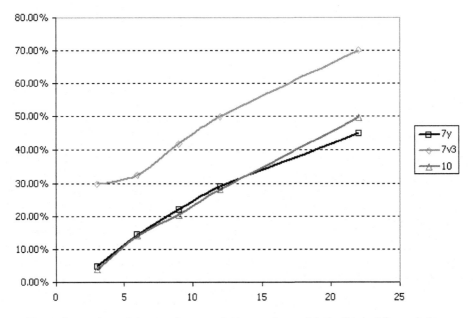

Fig. 6. Comparison of the spot 7 year and 10 year skews with the 7 into 3 forward skew.

5.1. The problem

In our economy, we assume that we have $(n+1)$ primary assets available for hedging with price processes $S^i = (S^i_t)_{t \geq 0}$. The first asset S^0 is the money market account, i.e., $S^0_t = \exp(\int_0^t r_s ds)$. It will be used as numeraire and all quantities will be expressed in units of S^0. We shall consider zero-coupon credit derivatives or contingent claims of the European type. The hedging asset S^i will represent the zero-coupon defaultable bond maturing at T linked to obligor i; i.e., it pays 1 if obligor i survives until time T, or 0 otherwise. Its payoff at maturity is defined as:

$$S^i_T \triangleq 1 - D^i_T.$$

In practice, zero-coupon defaultable bonds are rarely traded in the market. However, they can be recovered from the prices of standard default swaps with different maturities. A contingent claim is formally defined as follows.

Definition 5.1. A contingent claim is a $\mathcal{G}_T$-measurable random variable H_T describing the payoff at maturity T of a financial instrument.

This includes all types of portfolio credit derivatives.

Example 5.1. The payoff of a kth-to-default (zero-coupon note) maturing at T is defined as:

$$H_T^{(k)} \triangleq \mathbf{1}_{\left\{ \sum_{i=1}^n D^i_T < k \right\}}.$$

It pays 1 if there are less than k defaults in the basket, or 0 otherwise. The most common structure in this category is a first-to-default, $H_T^{(1)}$, which pays 1 if no obligor in the basket defaults before T.

Example 5.2. The payoff of a CDO tranche (zero-coupon note) covering the portfolio losses, which fall in the range $[K_1, K_2]$, where $0 \leq K_1 < K_2 \leq 1$, is

$$H_T^{(K_1,K_2)} \triangleq \frac{1}{K_2 - K_1} \min \left(\max \left(\frac{1}{n} \sum_{i=1}^n \left(1 - \delta^i\right) D_t^i - K_1, 0 \right), K_2 - K_1 \right).$$

We shall consider the problem of pricing and hedging zero-coupon contingent claims by dynamically trading the hedging assets S. We assume that P is a risk neutral martingale measure and we follow the approach of Föllmer and Sondermann [10], where a "good" martingale measure is chosen, then the minimization is done with respect to this measure.

As shown is Föllmer and Sondermann [10], the hedging strategy is obtained by the Kunita-Watanabe decomposition of the $\{\mathcal{G}_t\}$-martingale $H_t = \mathbb{E}[H_T | \mathcal{G}_t]$:

$$H_T = H_0 + \int_{]0,T]} (\alpha_t^{H_T})^{tr} dS_t + L_T^{H_T},$$

where L^{H_T} is a martingale orthogonal to S. Our goal is to find an analytical result for $(\alpha_t^{H_T})$.

5.2. *Marked point process*

We shall use the equivalent fatal shock description $(N^\pi)_{\pi \in \mathbf{\Pi_n}}$ to derive the Marked Point Process representation of the model.

We define the sequence of ordered default times $(T_0, T_1, \ldots, T_n)$:

$$T_0 = 0 \leq T_1 \leq \cdots \leq T_n,$$

and the identities of the defaulted obligors $(Z_0, Z_1, \ldots, Z_n)$ as:

$$T_0 = 0, Z_0 = \emptyset;$$
$$T_k = \min \{\tau_i : 1 \leq i \leq n, \tau_i > T_{k-1}\};$$
$$Z_k = \pi \text{ if } T_k = \tau_i \text{ for all } i \in \pi, \text{ and } \pi \in \mathbf{\Pi_n}.$$

The mark space of this point process is $E \triangleq \mathbf{\Pi_n}$, the set of all subsets of $I_n \triangleq \{1, \ldots, n\}$.

We denote by $(\lambda_t^\pi)_{t \geq 0}$ the $(P, \{\mathcal{G}_t\})$-intensity of the default time $\tau(\pi)$ when all the obligors in π default simultaneously:

$$\lambda_t^\pi = \left[\prod_{i \in \pi} \left(1 - D_t^i\right) \right] \times \left[\sum_{x \subset (I_n \setminus \pi)} \left[\prod_{i \in x} D_t^i \right] \lambda^{(\pi \cup x)} (X_t) \right].$$

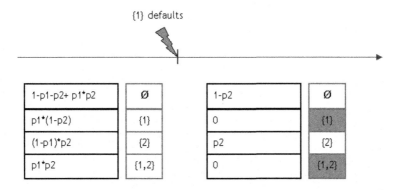

Fig. 7. Structure of the Market Point Process.

The double sequence $(T_k, Z_k)_{k \geq 1}$ defines a marked point process with counting measure

$$\mu(\omega, dt \times dz) : (\Omega, \mathcal{G}) \to ((0, \infty) \times E, (0, \infty) \otimes \mathcal{E}),$$

$$\int_0^t \int_E H(\omega, t, z) \mu(\omega, dt \times dz) = \sum_{k \geq 1} H(\omega, T_k(\omega), Z_k(\omega)) \mathbf{1}_{\{T_k(\omega) \leq t\}},$$

and $(P, \{\mathcal{G}_t\})$-intensity kernel $\lambda_t(\omega, dz) dt = \lambda_t(\omega) \phi_t(\omega, dz)$, which is given by

$$\lambda_t = \sum_{\pi \in \mathbf{\Pi_n}} \lambda_t^\pi,$$

$$\phi_t(\omega, \pi) = \frac{\lambda_t^\pi}{\lambda_t}, \text{ for } \pi \in \mathbf{\Pi_n},$$

with the convention $\phi_t(.) = 0$ if $\lambda_t = 0$. The pair $(\lambda_t, \phi_t(dz))$ is the $(P, \{\mathcal{G}_t\})$-local characteristics of the counting measure $\mu(dt \times dz)$.

5.3. Martingale representation

The agents' filtration $\{\mathcal{G}_t\}$ is generated by the Brownian motion W and the MPP $\mu(dt \times dz)$. We apply a martingale representation theorem (see [13]) to the $\{\mathcal{G}_t\}$-martingale $H_t = \mathbb{E}[H_T|\mathcal{G}_t]$.

Proposition 5.1. (*Martingale representation of H_t*). *The $\{\mathcal{G}_t\}$-martingale $H_t = \mathbb{E}[H_T|\mathcal{G}_t]$, $t \in [0, T^*]$, where H_T is a $\mathcal{G}_T$-measurable random variable, integrable with respect to P, admits the following integral representation*

$$H_t = H_0 + \int_0^t (\xi_s)^{tr} dW_s - \int_0^t \int_E \zeta(s, z) (\mu(ds \times dz) - \lambda_s(dz) ds),$$

where ξ is a d-dimensional $\{\mathcal{G}_t\}$-predictable process and $\zeta(s, z)$ is an E-indexed $\{\mathcal{G}_t\}$-predictable process such that

$$\int_0^t \|\xi_s\|^2 ds < \infty, \int_0^t \int_E \zeta(s, z) \lambda_s(dz) ds < \infty,$$

almost surely.

5.4. *Computing the hedging strategy*

We use the martingale representation of Proposition 5.1 to derive the risk-minimizing hedging strategy. This is equivalent to finding the Kunita–Watanabe decomposition of H_T:

$$H_T = H_0 + \int_{]0,T]} (\alpha_t)^{tr} dS_t + L_T.$$

Our goal is to establish an analytical result, which derives the single-name hedges $(\alpha^i)_{1 \le i \le n}$ in terms of the martingale representation predictable processes ξ and $\zeta(.,\pi)$, $\pi \in \mathbf{\Pi_n}$. As shown in Föllmer and Schweizer [11], the strategy (α) can be computed as

$$\alpha_t = d \langle S \rangle_t^{-1} d \langle S, V(\alpha) \rangle_t,$$

where the value process is given by

$$V_t(\alpha) = H_t = \mathbb{E}[H_T | \mathcal{G}_t], \quad \text{for } t \in [0,T].$$

This follows from the Kunita-Watanabe decomposition of H and the projection of $V_t(\alpha)$ on the martingale $\int_{]0,t]} (\alpha_s)^{tr} dS_s$.

Theorem 5.1. (*Risk-minimizing hedging strategy*). *The risk-minimizing hedging strategy of a general (basket) contingent claim with single-name instruments is given by the solution of the following linear system, for* $1 \le k \le n$,

$$\sum_{i=1}^n \alpha_t^i S_t^i \left[(\sigma_t^i)^{tr} \sigma_t^k + \int_E \mathbf{1}_{\{i \in z\}} \mathbf{1}_{\{k \in z\}} \lambda_t(dz) \right]$$

$$= (\sigma_t^k)^{tr} \xi_t + \int_E \zeta(t,z) \mathbf{1}_{\{k \in z\}} \lambda_t(dz),$$

where $(\sigma^i)_{1 \le i \le n}$ *are the volatilities of the single-name assets* $(S^i)_{1 \le i \le n}$.

Proof. The single-name dynamics are given by

$$(S_t^i) = \int_0^t S_{s-}^i \left((\sigma_s^i)^{tr} dW_s - \int_E \mathbf{1}_{\{i \in z\}} (\mu(ds \times dz) - \lambda_s(dz) ds) \right),$$

and the predictable covariance is

$$d \langle S \rangle_t^{i,j} = d \langle S^i, S^j \rangle_t = S_{t-}^i S_{t-}^j \left((\sigma_t^i)^{tr} \sigma_t^j + \int_E \mathbf{1}_{\{i \in z\}} \mathbf{1}_{\{j \in z\}} \lambda_t(dz) \right) dt.$$

The value process $V_t(\alpha) = \mathbb{E}[H_T | \mathcal{G}_t]$ is given by the martingale representation

$$V_t(\alpha) = \mathbb{E}[H_T | \mathcal{G}_t] = H_0 + \int_0^t (\xi_s)^{tr} dW_s - \int_0^t \int_E \zeta(s,z) (\mu(ds \times dz) - \lambda_s(dz) ds).$$

Hence, we have

$$d \langle S, V(\alpha) \rangle_t^i = S_{t-}^i \left((\sigma_t^i)^{tr} \xi_t + \int_E \mathbf{1}_{\{i \in z\}} \zeta(t,z) \lambda_t(dz) \right) dt. \qquad \square$$

If we assume that the dynamics of the state-variables' vector are given by the following SDE:

$$dX_t = \alpha\left(X_t\right)dt + \beta\left(X_t\right)dW_t,$$

for some Lipschitz functions $\alpha_k : \mathbb{R}^d \to \mathbb{R}$ and $\beta_{kj} : \mathbb{R}^d \to \mathbb{R}^d$, $1 \leq k, j \leq d$, then by applying Itô's lemma and using the Markovian property of X, we can write explicitly the dynamics of the single-name instruments $(S^i)_{1 \leq i \leq n}$:

$$S_t^i = \left(1 - D_t^i\right)\mathbb{E}\left[\exp\left(-\int_t^T \lambda^i\left(X_s\right)ds\right) | \mathcal{F}_t\right], \quad \text{for } 1 \leq i \leq n.$$

Lemma 5.1. (*Single-name dynamics*). *We have*

$$S_t^i = S_0^i - \int_0^t \int_E s^i\left(s, X_s\right)\mathbf{1}_{\{i \in z\}}\left(\mu\left(ds \times dz\right) - \lambda_s\left(dz\right)ds\right)$$

$$+ \int_0^t \left(1 - D_s^i\right)\sum_{j=1}^d \sum_{k=1}^d \frac{\partial s^i\left(s, X_s\right)}{\partial x_j}\beta_{jk}\left(X_s\right)dW_s^k,$$

where $s^i(t, x) : [0, T] \times \mathbb{R}^d \to \mathbb{R}$ *is defined as*

$$s^i\left(t, x\right) \triangleq \mathbb{E}_{(t,x)}\left[\exp\left(-\int_t^T \lambda^i\left(X_s\right)ds\right)\right].$$

Lemma 5.1 establishes the martingale representation for the single-name securities whose payoff is $H_T = 1 - D_T^i$:

$$\zeta^{1-D_T^i}\left(t, z\right) = \mathbf{1}_{\{i \in z\}}s^i\left(t, X_t\right), \quad \text{for } z \in \mathbf{\Pi_n},$$

$$\left(\xi_t^{1-D_T^i}\right)^k = \left(1 - D_t^i\right)\sum_{j=1}^d \frac{\partial s^i\left(t, X_t\right)}{\partial x_j}\beta_{jk}\left(X_t\right), \quad \text{for } 1 \leq k \leq d.$$

The process $\zeta^{1-D_T^i}(t, z)$ represents the jump-to-default risk, and $(\xi_t^{1-D_T^i})^k$ is the delta with respect to each factor W^k.

Thus, the hedging strategy is the solution of the following linear system, for $1 \leq k \leq n$,

$$\sum_{i=1}^n \alpha_t^i\left[\int_E \zeta^{1-D_T^k}\left(t, z\right)\zeta^{1-D_T^i}\left(t, z\right)\lambda_t\left(dz\right) + \left(\xi_t^{1-D_T^k}\right)^{tr}\xi_t^{1-D_T^i}\right]$$

$$= \int_E \zeta\left(t, z\right)\zeta^{1-D_T^k}\left(t, z\right)\lambda_t\left(dz\right) + \left(\xi_t^{1-D_T^k}\right)^{tr}\xi_t.$$

Note that this problem combines both default risk and spread risk.

To illustrate the method, we show how the calculations are performed for a first-to-default basket.

Example 5.3. The payoff of a first-to-default contingent claim is given by

$$H_T^{(1)} \triangleq \prod_{i=1}^{n} \left(1 - D_T^i\right).$$

The price of this claim at time t is

$$H_t^{(1)} \triangleq \mathbb{E}\left[\prod_{i=1}^{n}\left(1 - D_T^i\right) | \mathcal{G}_t\right].$$

We can show that it can be expressed as

$$H_t^{(1)} \triangleq \left[\prod_{i=1}^{n}\left(1 - D_t^i\right)\right] h^{(1)}\left(t, X_t\right),$$

where the function $h^{(1)}(t, x) : [0, T] \times \mathbb{R}^d \to \mathbb{R}$ is defined as

$$h^{(1)}\left(t, x\right) = \mathbb{E}_{(t,x)}\left[\exp\left(-\int_t^T \lambda^{(1)}\left(X_s\right) ds\right)\right],$$

$$\lambda^{(1)}\left(X_t\right) = \sum_{j=1}^{m}\left[1 - \prod_{i=1}^{n}\left(1 - p^{i,j}\right)\right]\lambda^{c_j}\left(X_t\right).$$

Using Itô's lemma and some algebra, we find

$$dH_t^{(1)} = -\int_E h^{(1)}\left(t, X_t\right)\left(\mu\left(dt \times dz\right) - \lambda_t\left(dz\right) dt\right)$$

$$+ \left[\prod_{i=1}^{n}\left(1 - D_t^i\right)\right]\sum_{j=1}^{d}\sum_{k=1}^{d}\frac{\partial h^{(1)}\left(t, X_t\right)}{\partial x_j}\beta_{jk}\left(X_t\right) dW_t^k.$$

This gives the processes of the martingale representation

$$\zeta^{H_T^{(1)}}\left(t, z\right) = h^{(1)}\left(t, X_t\right), \quad \text{for all } z \in \mathbf{\Pi_n},$$

$$\left(\xi_t^{H_T^{(1)}}\right)^k = \left[\prod_{i=1}^{n}\left(1 - D_t^i\right)\right]\sum_{j=1}^{d}\frac{\partial h^{(1)}\left(t, X_t\right)}{\partial x_j}\beta_{jk}\left(X_t\right), \quad \text{for } 1 \leq k \leq d,$$

which can be plugged into the linear system of Theorem 5.1. Inverting this latter gives the single-name hedge ratios of the first-to-default basket claim.

6. Conclusion

In this paper, we have presented a dynamic version of the Marshall–Olkin model. The basic motivation of our approach was to build a dynamic top-down model, which is consistent with single-name information. By using Marshall–Olkin as a starting point, we are able to achieve tractability while preserving the key features of the original framework. The numerical implementation is done analytically. After surveying the various approximations in the literature, we have derived an exact closed form solution of the loss density, which is based on an asymptotic series expansion. Our formula is applicable to all intensity processes whose Laplace transform is known analytically.

Furthermore, building up on the time-invariance property of the multivariate Poisson process, we have investigated the implied forward skew as a natural application of the model. Indeed, given the memory-less property of a Poisson process, the forward-looking stochastic environment, in a Marshall–Olkin framework, is naturally well behaved. And the forward skew fits in quite well. This is not the case with most copula models.

Finally, no dynamic model would be complete without the supporting dynamic hedging strategies *a la* Black-Scholes. With this in mind, we have developed a self-contained theory of dynamic hedging within this framework. We have addressed the issue of market incompleteness by using the quadratic risk minimization approach of Föllmer-Sonderman.

References

[1] R. Barlow and F. Proschan, *Statistical Theory of Reliability and Life Testing* (Silver Spring, Maryland, 1981).

[2] N. Bennani, The forward loss model: A dynamic term structure approach for the pricing of portfolio credit derivatives, Working Paper (2005).

[3] P. Brémaud, *Point Processes and Queues: Martingale Dynamics* (Springer-Verlag, New York, 1980).

[4] D. Brigo, A. Pallavicini and R. Torresetti, Calibration of cdo tranches with the dynamical generalized poisson loss model, Working Paper (2006).

[5] E. Errais, K. Giesecke, L. Goldberg, Pricing credit from the top down with affine point processes, Working Paper (2006).

[6] D. Duffie, First-To-Default valuation, Working Paper, Graduate School of Business, Stanford University (1998).

[7] D. Duffie and N. Garleanu, Risk and valuation of collateralized debt obligations, *Financial Analysts Journal* **57**(1) (2001) 41–59.

[8] D. Duffie and J. Pan, Analytical value-at-risk with jumps and credit risk, *Finance and Stochastics* **5** (2001) 155–180.

[9] D. Duffie and K. Singleton, Simulating correlated defaults, Working Paper, Graduate School of Business, Stanford University (1999).

[10] H. Föllmer and D. Sondermann, Heding of non-redundant contingent claims, in W. Hildenbrand and A. Mas-Colell eds. *Contributions to Mathematical Economics* (North-Holland, Amsterdam, 1986) pp. 205–223.

[11] H. Föllmer and M. Schweizer, Hedging of contingent claims under incomplete information, in M. H. A. Davis, and R. J. Elliott eds., *Applied Stochastic Analysis* (Gordon and Breach, London, 1991), pp. 389–414.

[12] K. Giesecke and L. Goldberg, A top down approach to multi-name credit, Working Paper (2005).

[13] J. Jacod and A. Shiryaev, *Limit Theorems for Stochastic Processes* (Springer-Verlag, New York, 1987).

[14] F. Lindskog and A. McNeil, Common poisson shock models: Applications to insurance and credit risk modelling, *ASTIN Bulletin* **33**(2) (2003) 209–238.

[15] F. Longstaff and A. Rajan, An empirical analysis of the piricing of collateralized debt obligations, Working Paper (2006).

[16] A. Patton, Modelling Time-varying exchange rate dependence using the conditional copula, Working Paper 2001-09, University of California, San Diego (2001).

[17] P. Protter, *Stochastic Integration and Differential Equations*, Second Edition, Version 2.1 (Springer-Verlag, New York, 2005).

[18] P. Schönbucher, Portfolio losses and the term structure of loss transition rates: A new methodology for the pricing of portfolio credit derivatives, Working Paper (2005).

[19] J. Sidenius, V. Piterbag and L. Andersen, A new framework for dynamic credit portfolio loss modeling, Working Paper (2005).

[20] A. Sklar, Fonctions de répartition à n dimensions et leurs marges, *Publ. Inst. Statist. Univ. Paris* **8** (1959) 229–231.

Appendix A. Proof of Technical Results

In this appendix, we give the proofs of some of the technical results in the paper.

A.1. *Proof of Lemma 3.1*

Proof. We proceed by induction. We assume that the property is verified for n, and we prove that it holds for $n + 1$.

For $n + 1$ names, the Fourier transform of the loss variable L_T^{n+1}, defined as

$$L_T^{n+1} \triangleq \sum_{i=1}^{n+1} L_i D_T^i, \tag{A.1}$$

is given by

$$\phi^{n+1}(u) = \mathbb{E}\left[\exp\left(-iuL_T^{n+1}\right)\right].$$

Conditioning on D_T^{n+1}, we have

$$\phi^{n+1}(u) = \mathbb{E}\left[\mathbb{E}\left[\exp\left(-iuL_T^{n+1}\right)\big|D_T^{n+1}\right]\right]$$

$$= \mathbb{E}\left[\exp\left(-iuL^{n+1}D_T^{n+1}\right)\mathbb{E}\left[\exp\left(-iuL_T^n\right)\big|D_T^{n+1}\right]\right]. \tag{A.2}$$

Using the induction relationship, we can write the conditional characteristic function as

$$\mathbb{E}[\exp(-iuL_T^n)|D_T^{n+1}] = \sum_{\pi_n \in \mathbf{\Pi_n}} \mathbb{E}\left(\prod_{i\in\pi_n}\left(1 - D_T^i\right)\big|D_T^{n+1}\right)\left[\prod_{i\notin\pi_n}\psi_i \prod_{i\in\pi_n}\left(1 - \psi_i\right)\right],$$

and Eq. (A.2) becomes

$$\phi^{n+1}(u) = \mathbb{E}\left[\psi_{n+1}^{D_T^{n+1}}\left[\sum_{\pi_n \in \mathbf{\Pi_n}} \mathbb{E}\left(\prod_{i\in\pi_n}\left(1 - D_T^i\right)\big|D_T^{n+1}\right)\left[\prod_{i\notin\pi_n}\psi_i \prod_{i\in\pi_n}\left(1 - \psi_i\right)\right]\right]\right].$$

Writing the expectation explicitly, we obtain

$$
\phi^{n+1}(u) = \mathbb{P}\left(D_T^{n+1} = 0\right) \sum_{\pi_n \in \mathbf{\Pi_n}} \mathbb{E}\left(\prod_{i \in \pi_n} \left(1 - D_T^i\right) | \{D_T^{n+1} = 0\}\right)
$$

$$
\times \left[\prod_{i \notin \pi_n} \psi_i \prod_{i \in \pi_n} \left(1 - \psi_i\right)\right]
$$

$$
+ \mathbb{P}\left(D_T^{n+1} = 1\right) \psi_{n+1} \sum_{\pi_n \in \mathbf{\Pi_n}} \mathbb{E}\left(\prod_{i \in \pi_n} \left(1 - D_T^i\right) | \{D_T^{n+1} = 1\}\right)
$$

$$
\times \left[\prod_{i \notin \pi_n} \psi_i \prod_{i \in \pi_n} \left(1 - \psi_i\right)\right]. \tag{A.3}
$$

Observing that

$$
\mathbb{P}\left(D_T^{n+1} = 0\right) \mathbb{E}\left(\prod_{i \in \pi_n} \left(1 - D_T^i\right) | \{D_T^{n+1} = 0\}\right)
$$

$$
= \mathbb{E}\left(\left[\prod_{i \in \pi_n} \left(1 - D_T^i\right)\right] \left(1 - D_T^{n+1}\right)\right),
$$

and

$$
\mathbb{P}\left(D_T^{n+1} = 1\right) \mathbb{E}\left(\prod_{i \in \pi_n} \left(1 - D_T^i\right) | \{D_T^{n+1} = 1\}\right)
$$

$$
= \mathbb{E}\left(\left[\prod_{i \in \pi_n} \left(1 - D_T^i\right)\right] D_T^{n+1}\right)
$$

$$
= \mathbb{E}\left(\left[\prod_{i \in \pi_n} \left(1 - D_T^i\right)\right]\right) - \mathbb{E}\left(\left[\prod_{i \in \pi_n} \left(1 - D_T^i\right)\right] \left(1 - D_T^{n+1}\right)\right),
$$

Eq. (A.3) becomes

$$
\phi^{n+1}(u) = \sum_{\pi_n \in \mathbf{\Pi_n}} \mathbb{E}\left(\left[\prod_{i \in \pi_n} \left(1 - D_T^i\right)\right] \left(1 - D_T^{n+1}\right)\right) \left(1 - \psi_{n+1}\right)
$$

$$
\times \left[\prod_{i \notin \pi_n} \psi_i \prod_{i \in \pi_n} \left(1 - \psi_i\right)\right] + \sum_{\pi_n \in \mathbf{\Pi_n}} \mathbb{E}\left(\left[\prod_{i \in \pi_n} \left(1 - D_T^i\right)\right]\right) \psi_{n+1}
$$

$$
\times \left[\prod_{i \notin \pi_n} \psi_i \prod_{i \in \pi_n} \left(1 - \psi_i\right)\right]. \tag{A.4}
$$

The set of subsets $\mathbf{\Pi_{n+1}}$ is partitioned into two sets $\mathbf{\Pi_{n+1}^+}$ and $\mathbf{\Pi_{n+1}^-}$:

$$\mathbf{\Pi_{n+1}^+} = \{\pi_{n+1} : \pi_{n+1} \in \mathbf{\Pi_{n+1}}, (n+1) \in \pi_{n+1}\},$$

$$\mathbf{\Pi_{n+1}^-} = \{\pi_{n+1} : \pi_{n+1} \in \mathbf{\Pi_{n+1}}, (n+1) \notin \pi_{n+1}\}.$$

For $\pi_{n+1} \in \mathbf{\Pi_{n+1}^+}$: $\pi_{n+1} = \pi_n \cup \{n+1\}$, we have

$$Q_{\pi_{n+1}}^{[1]}(T) = \mathbb{E}\left(\prod_{i \in \pi_{n+1}}\left(1 - D_T^i\right)\right) = \mathbb{E}\left(\left[\prod_{i \in \pi_n}\left(1 - D_T^i\right)\right] \cdot \left(1 - D_T^{n+1}\right)\right),$$

and

$$n + 1 \in \pi_{n+1}.$$

For $\pi_{n+1} \in \mathbf{\Pi_{n+1}^-}$: $\pi_{n+1} = \pi_n \cup \emptyset$, we have

$$Q_{\pi_{n+1}}^{[1]}(T) = \mathbb{E}\left(\prod_{i \in \pi_{n+1}}\left(1 - D_T^i\right)\right) = \mathbb{E}\left(\prod_{i \in \pi_n}\left(1 - D_T^i\right)\right),$$

and

$$n + 1 \notin \pi_{n+1}.$$

Substituting in Eq. (A.4) yields

$$\phi^{n+1}(u) = \sum_{\pi_{n+1} \in \mathbf{\Pi_{n+1}^+}} Q_{\pi_{n+1}}^{[1]}(T)\left[\prod_{i \notin \pi_{n+1}} \psi_i \prod_{i \in \pi_{n+1}}\left(1 - \psi_i\right)\right]$$

$$+ \sum_{\pi_{n+1} \in \mathbf{\Pi_{n+1}^-}} Q_{\pi_{n+1}}^{[1]}(T)\left[\prod_{i \notin \pi_{n+1}} \psi_i \prod_{i \in \pi_{n+1}}\left(1 - \psi_i\right)\right],$$

which ends the proof. $\qquad\qquad\qquad\qquad\qquad\qquad\qquad\qquad\qquad\qquad\square$

A.2. *Proof of Theorem 3.1*

Proof. Using Lemma 3.1, we express the probability generating function of the random variable L_T in terms of the FTD Q-factors $Q_\pi^{[1]}(T)$

$$\phi(u) = \sum_{\pi \in \mathbf{\Pi_n}} Q_\pi^{[1]}(T)\left[\prod_{i \notin \pi} \psi_i \prod_{i \in \pi}\left(1 - \psi_i\right)\right]. \tag{A.5}$$

Each FTD Q-factor $Q_\pi^{[1]}(T)$ is given by

$$Q_\pi^{[1]}(T) = \mathbb{E}\left[\exp\left(-\left[\sum_{j=1}^{m_c}\left(1 - \prod_{i \in \pi}\left(1 - p^{i,j}\right)\right)\Lambda_T^{c_j}\right] - \left[\sum_{i \in \pi}\Lambda_T^{0,i}\right]\right)\right]. \tag{A.6}$$

We expand the exponentials

$$\exp\left(\prod_{i\in\pi}\left(1-p^{i,j}\right)\Lambda_T^{c_j}\right)=\sum_{n_j=0}^{+\infty}\frac{\left[\prod_{i\in\pi}\left(1-p^{i,j}\right)\Lambda_T^{c_j}\right]^{n_j}}{n_j!}.$$

We substitute in Eq. (A.6)

$$Q_\pi^{[1]}\left(T\right)=\mathbb{E}\left[e^{\left[-\sum_{i\in\pi}\Lambda_T^{0,i}\right]}\sum_{n_1=0}^{+\infty}\cdots\sum_{n_m=0}^{+\infty}e^{-\Lambda_T^{c_1}}\frac{\left[\prod_{i\in\pi}\left(1-p^{i,1}\right)\Lambda_T^{c_1}\right]^{n_1}}{n_1!}\right.$$

$$\left.\cdots e^{-\Lambda_T^{c_m}}\frac{\left[\prod_{i\in\pi}\left(1-p^{i,m_c}\right)\Lambda_T^{c_{m_c}}\right]^{n_{m_c}}}{n_{m_c}!}\right]$$

$$=\mathbb{E}\left[\sum_{n_1=0}^{+\infty}\cdots\sum_{n_{m_c}=0}^{+\infty}\left[\prod_{j=1}^{m_c}e^{-\Lambda_T^{c_j}}\frac{\left(\Lambda_T^{c_j}\right)^{n_j}}{n_j!}\right]\right.$$

$$\left.\times e^{-\sum_{i\in\pi}\left(\Lambda_T^{0,i}+\log\left(\left(1-p^{i,1}\right)^{n_1}\right)+\cdots+\log\left(\left(1-p^{i,m_c}\right)^{n_{m_c}}\right)\right)}\right]$$

$$=\sum_{n_1=0}^{+\infty}\cdots\sum_{n_{m_c}=0}^{+\infty}\left[\prod_{j=1}^{m_c}\mathbb{E}\left[e^{-\Lambda_T^{c_j}}\frac{\left(\Lambda_T^{c_j}\right)^{n_j}}{n_j!}\right]\right]$$

$$\times\mathbb{E}\left[e^{-\sum_{i\in\pi}\left(\Lambda_T^{0,i}+\log\left(\left(1-p^{i,1}\right)^{n_1}\right)+\cdots+\log\left(\left(1-p^{i,m_c}\right)^{n_{m_c}}\right)\right)}\right].$$

In the last equality, we have used the fact that the intensity processes $(\Lambda_T^{c_1},\ldots\Lambda_T^{c_{m_c}},\Lambda_T^{0,1},\ldots,\Lambda_T^{0,n})$ are independent. Substituting in the expression of the characteristic function yields

$$\phi(u)=\sum_{n_1=0}^{+\infty}\cdots\sum_{n_{m_c}=0}^{+\infty}\left[\prod_{j=1}^{m_c}\mathbb{E}\left[e^{-\Lambda_T^{c_j}}\frac{\left(\Lambda_T^{c_j}\right)^{n_j}}{n_j!}\right]\right]$$

$$\times\sum_{\pi\in\mathbf{\Pi_n}}Q_{n_1,\ldots,n_{m_c}}\left(\pi\right)\left[\prod_{i\notin\pi}\psi_i\prod_{i\in\pi}\left(1-\psi_i\right)\right],$$

where

$$Q_{n_1,\ldots,n_{m_c}}\left(\pi\right)\triangleq\mathbb{E}\left[\exp\left(-\sum_{i\in\pi}\left(\Lambda_T^{0,i}+\log\left(\left(1-p^{i,1}\right)^{n_1}\right)+\cdots\right.\right.\right.$$

$$\left.\left.\left.+\log\left(\left(1-p^{i,m_c}\right)^{n_{m_c}}\right)\right)\right)\right].$$

We define the function $\phi_{n_1,\ldots,n_{m_c}}\left(u\right)$

$$\phi_{n_1,\ldots,n_{m_c}}\left(u\right)\triangleq\sum_{\pi\in\mathbf{\Pi_n}}Q_{n_1,\ldots,n_{m_c}}\left(\pi\right)\left[\prod_{i\notin\pi}\psi_i\prod_{i\in\pi}\left(1-\psi_i\right)\right].\tag{A.7}$$

By virtue of Lemma 3.1, $\phi_{n_1,\dots,n_{m_c}}(u)$ is the Fourier transform of n independent idiosyncratic default variables with intensities $(\Lambda_T^i(n_1,\dots,n_{m_c}))_{1 \le i \le n}$

$$\Lambda_T^i(n_1,\dots,n_{m_c}) \triangleq \Lambda_T^{0,i} + \log\left(\left(1-p^{i,1}\right)^{n_1}\right) + \cdots + \log\left(\left(1-p^{i,m_c}\right)^{n_{m_c}}\right).$$

Hence

$$\phi_{n_1,\dots,n_{m_c}}(u) = \prod_{i=1}^{n}\left(\psi_i + \left(\mathbb{E}\left[e^{-\Lambda_T^{0,i}}\right]\left(1-p^{i,1}\right)^{n_1}\dots\left(1-p^{i,m_c}\right)^{n_{m_c}}\right)(1-\psi_i)\right).$$

In other words, $\phi_{n_1,\dots,n_{m_c}}(u)$ is the Fourier transform of the random variable $\sum_{i=1}^{n} Y_i$, where $(Y_1,\dots,Y_n)$ are n independent Bernoulli variables with parameters $p_{n_1,\dots,n_{m_c}} = (p_{n_1,\dots,n_{m_c}}(1),\dots,p_{n_1,\dots,n_{m_c}}(n))$

$$p_{n_1,\dots,n_{m_c}}(i) \triangleq 1 - \mathbb{E}\left[e^{-\Lambda_T^{0,i}}\right]\left(1-p^{i,1}\right)^{n_1}\dots\left(1-p^{i,m_c}\right)^{n_{m_c}}.$$

Thus, $\phi(u)$ is a weighted average of the conditional independent characteristic functions $\phi_{n_1,\dots,n_{m_c}}(u)$

$$\phi(u) = \sum_{n_1=0}^{+\infty}\cdots\sum_{n_{m_c}=0}^{+\infty} w_{n_1,\dots,n_{m_c}} \cdot \phi_{n_1,\dots,n_{m_c}}(u),$$

$$w_{n_1,\dots,n_{m_c}} = \mathbb{E}\left[e^{-\Lambda_T^{c_1}}\frac{(\Lambda_T^{c_1})^{n_1}}{n_1!}\right]\dots\mathbb{E}\left[e^{-\Lambda_T^{c_{m_c}}}\frac{(\Lambda_T^{c_{m_c}})^{n_{m_c}}}{n_{m_c}!}\right]. \tag{A.8}$$

Inverting the Fourier transform $\phi(u)$ concludes the proof. $\qquad\square$

JOINT DISTRIBUTIONS OF PORTFOLIO LOSSES AND EXOTIC PORTFOLIO PRODUCTS

FRIEDEL EPPLE, SAM MORGAN and LUTZ SCHLOEGL*

Lehman Brothers, 25 Bank Street, London E14 5LE, United Kingdom
*luschloe@lehman.com

The pricing of exotic portfolio products, e.g. path-dependent CDO tranches, relies on the joint probability distribution of portfolio losses at different time horizons. We discuss a range of methods to construct the joint distribution in a way that is consistent with market prices of vanilla CDO tranches. As an example, we show how our loss-linking methods provide estimates for the breakeven spreads of forward-starting tranches.[1]

Keywords: CDOs; correlation modelling; path-dependent portfolio derivatives.

1. Introduction: Pricing Exotic Portfolio Products

With the establishment of a liquid index tranche market, implied default correlation is increasingly perceived as a market observable. The current market standard for pricing vanilla CDOs is the one-factor Gaussian copula model in conjunction with base correlation. With some interpolation assumptions, the base correlation model can be used to back out portfolio loss distributions at different time horizons. Also, using mapping procedures, we can imply loss distributions for bespoke CDO tranches and CDO^2 structures. For example, Fig. 1 shows the loss distributions for the particular bespoke portfolio used in this article at 5 and 10 year time horizons as of 30 May 2006.[2]

As the credit correlation market evolves, more exotic structures are gaining in popularity. Among these, we find instruments where value depends on the distribution of incremental losses (e.g. forward-starting CDOs) and fully path-dependent structures such as reserve-account CDOs. To price these, knowledge of the implied loss distributions to various horizons is insufficient. Instead, we need to know the joint distribution of losses at different maturities. Therefore, we must join

---

*Corresponding author.

[1]The authors would like to thank Matthew Livesey for his collaboration on part of this work.

[2]The portfolio consists of a pool of 160 equally-weighted European credits with an average 10 year spread of 63 basis points. The standard deviation of the spread distribution is 34 basis points.

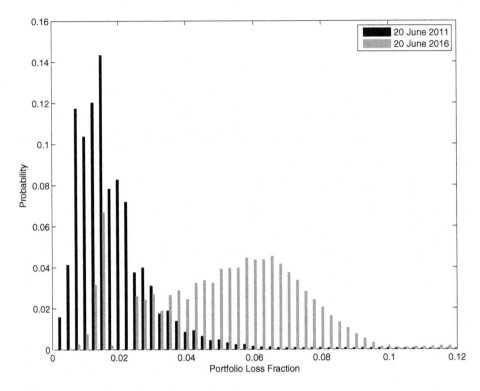

Fig. 1. Marginal loss distributions implied by base correlation for a bespoke portfolio. *Source*: Lehman Brothers.

the marginal loss distributions in a consistent and arbitrage-free way. The vanilla tranche market tells us very little about how this should be done.

In this article we discuss the constraints we face when building a joint loss distribution from marginal loss distributions at different maturities and describe various methods to link loss distributions consistently. We also compare the different methods, in particular looking at the breakeven spreads they give for forward-starting CDOs.

1.1. *Modelling framework*

We assume that the cumulative portfolio loss L_t is defined on a discrete support of size N consisting of loss levels K_i, $i = 0, \ldots, N - 1$. In general, there is no requirement that $K_0 = 0$ or that the losses are evenly spaced, although in practice both of these conditions are usually met. The basic modelling problem can then be described as follows: Given marginal loss distributions at times t_1 and $t_2 > t_1$,

$$p_i(t_1) := \mathrm{P}[L_{t_1} = K_i], \quad p_j(t_2) := \mathrm{P}[L_{t_2} = K_j], \tag{1.1}$$

we want to construct a consistent arbitrage-free joint distribution

$$p_{ij}(t_1, t_2) := \mathrm{P}[L_{t_1} = K_i,\, L_{t_2} = K_j]. \tag{1.2}$$

For simplicity, we will generally drop the time arguments when there is no ambiguity.

The matrix p_{ij} is constrained by the fact that the rows and columns must add up to the respective marginal distributions. Also, as the loss process is non-decreasing, p_{ij} is upper-triangular; we will refer to this as the monotonicity constraint. So, in summary, we have:

$$\sum_j p_{ij} = p_i(t_1), \quad \sum_i p_{ij} = p_j(t_2); \quad p_{ij} = 0, \quad \forall\, i > j. \tag{1.3}$$

In addition to these constraints we must also ensure that $0 \le p_{ij} \le 1$.

If there were no overlap between the loss distributions at t_1 and t_2, i.e., if the maximum loss at t_1 were less than or equal to the minimum loss at t_2, the monotonicity requirement would not impose an additional constraint. In particular, it would be possible to have independence between the losses at t_1 and t_2 in this case. In practice however, the loss distributions will always overlap. As the overlap between the supports of the two distributions grows, the constraints from the monotonicity condition become increasingly severe. From a pricing perspective, severe constraints are attractive as they reduce the pricing uncertainty: observations of the marginal distributions allow us to learn more about the joint distribution than would otherwise be possible.

2. Joining Loss Distributions Across Different Maturities

The constraints in Eq. (1.3) do not uniquely determine the joint loss distribution. In general, an infinite number of joint loss distributions are consistent with a given pair of marginal distributions. We can therefore follow a number of different approaches to link the marginals into a joint distribution. These approaches fall into two broad categories. On the one hand, we can attempt to calibrate a skew-consistent, arbitrage-free model of the portfolio loss to the vanilla CDO market and back out the implied joint distribution. This is currently a very active area of research, and many different models have been proposed, with varying degrees of success. In Sec. 2.4 we show results from a simple implementation of Schönbucher's forward loss rate model [3].

On the other hand, we can take a more model-independent view by using the marginals as inputs and imposing a dependence structure exogenously. Although the first approach clearly has more potential in terms of developing a unified framework for pricing portfolio exotics, the second allows us to focus on the effect of the copula joining the distributions across time. It will also prove useful as a way of obtaining reference points for breakeven spreads. We discuss several methods in this category next.

2.1. *The comonotonic method*

The comonotonic method is designed to introduce the maximum positive dependence between the losses at two time horizons consistent with a given set of marginal distributions. The method is developed from the well-known fact that, given a cumulative loss distribution F_t at some time t,

$$F_t(x) := P[L_t \leq x], \tag{2.1}$$

we can sample from this distribution by generating a uniform random variate U and solving for the value of i such that

$$F_t(K_{i-1}) < U \leq F_t(K_i). \tag{2.2}$$

This approach can be extended to generate a path for the cumulative loss through time by using the same value of U to sample from the loss distributions at each time horizon. If the marginal distributions are compatible with an increasing process, meaning that they satisfy

$$F_{t_1}(K_i) \geq F_{t_2}(K_i), \qquad \forall\, K_i, \quad \forall t_1 < t_2, \tag{2.3}$$

the resulting loss path is non-decreasing. If the marginal distributions are continuous, this approach generates losses L_{t_1} and L_{t_2} given by

$$L_{t_1} = F_{t_1}^{-1}(U), \quad L_{t_2} = F_{t_2}^{-1}(U) = F_{t_2}^{-1}(F_{t_1}(L_{t_1})). \tag{2.4}$$

Hence, the loss at time t_2 is uniquely determined by the loss at t_1 and this method generates the maximum positive dependence between the two losses. If the marginal distributions are discrete, conditioning on a particular loss at t_1 is equivalent to conditioning on a range of values for U and there is a corresponding range of values for the loss at t_2. Integrating over U we obtain the distribution for the loss at t_2 conditional on the loss at t_1.

While the comonotonic method is easy to implement, it implies rather unrealistic dynamics for the underlying loss process as the loss at any given time completely determines the loss at all future maturities up to discretisation effects. However, since the comonotonic method displays the maximum positive dependence, it at least represents a useful benchmark for pricing path-dependent exotics.

To visualize joint distributions, we use so-called heat maps for which the joint distribution is plotted in a two-dimensional chart with different shades of grey representing the magnitude of the probability at different points in the two-dimensional loss grid. We use a base-10 logarithmic scale with dark shades indicating a high probability. The losses at t_1 and t_2 are represented along the vertical and horizontal axes, respectively.

A heat map for the comonotonic method is shown in Fig. 2. The marginal distributions used in the construction of the joint distribution are the same as in Fig. 1. With the comonotonic method, the joint distribution takes a very simple form as the entire probability mass is concentrated along a curve in the upper right half of the loss grid. This reflects the fact that the loss at t_2 is almost uniquely determined by the loss at t_1.

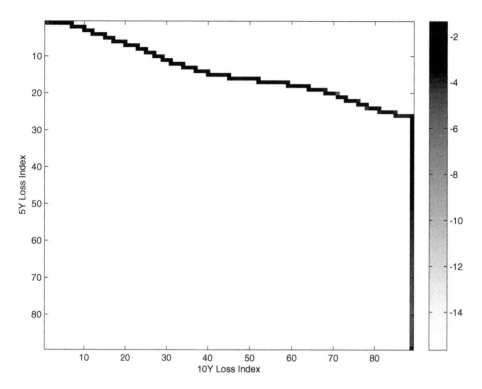

Fig. 2. Joint loss distribution for the comonotonic method. Losses at 5 years and 10 years are represented along the vertical and horizontal axes respectively. Dark shades represent high probability. *Source*: Lehman Brothers.

2.2. *Independent increments*

In a first attempt to build a more realistic joint loss distribution, we try to introduce independent increments, i.e., we assume that the incremental loss $\Delta L(t_1, t_2) := L_{t_2} - L_{t_1}$ between t_1 and t_2 is independent of the loss at t_1. From a modeling perspective, this is particularly attractive as it means that, observing losses in the near future, we do not learn anything about losses occurring later. However, as we will see, the constraints of Eq. (1.3) imposed on the joint distribution along with the requirement $0 \leq p_{ij} \leq 1$ usually rule out independent increments.

It is clear that independent increments are not always possible because they are usually inconsistent with the existence of a maximum loss in the portfolio. As a simple example, consider the case in which there is a non-zero probability of obtaining the maximum loss at t_1. The incremental loss conditional on having the maximum loss at t_1 must be deterministically zero. But if the loss increment is independent of the initial loss, then this same distribution applies to all losses at time t_1 and we conclude that the marginal distributions at t_1 and t_2 are identical, which of course is not generally the case.

Nonetheless, there are situations where an independent loss increment is approximately possible and it is instructive to consider this approach to see how it fails. To construct a joint distribution using independent increments, we have to find probabilities

$$q_n := \mathrm{P}[\Delta L(t_1, t_2) = K_n], \tag{2.5}$$

such that we recover the loss distribution at t_2 by carrying out the convolution of the loss distribution at t_1 with the incremental loss distribution given by the values of q_n:

$$p_n(t_2) = \sum_{k=0}^{n} p_k(t_1) q_{n-k}, \quad \forall\, n \in \{0, \ldots, N-1\}. \tag{2.6}$$

Based on this relation, we can easily determine the values of q_n via a bootstrapping procedure. However, it is important to note that the numbers obtained in this way are not guaranteed to take values in the interval $[0, 1]$ and they do not necessarily sum to unity. Hence, the bootstrap can easily lead to inconsistent results. We found that for short initial maturities t_1, the independent increment method works fairly well. For longer initial maturities, the bootstrap procedure becomes unstable and typically leads to values of q_n outside the allowed range of $[0, 1]$ after a few steps. This can be explained by the fact that the bootstrapping equation can be written in the form

$$q_n - q_{n-1} = \frac{1}{p_0(t_1)} \left(p_n(t_2) - p_{n-1}(t_2) - \sum_{k=0}^{n-2} q_k \big(p_{n-k}(t_1) - p_{n-k-1}(t_1) \big) \right.$$
$$\left. - q_{n-1} p_1(t_1) \right), \tag{2.7}$$

which means that any increments in the marginals are magnified by a factor $[p_0(t_1)]^{-1}$. For long initial maturities, the probability of a zero loss approaches zero, rendering the bootstrap unstable.

2.3. *The maximum entropy method*

The failure of independent increments to provide a reliable method to join marginal distributions can be stated as follows: Given a pair of marginal loss distributions, the non-decreasing property of the loss process may require a dependence structure that allows us to gain some information about the incremental loss $\Delta L(t_1, t_2)$ by observing the initial loss L_{t_1}. As we cannot eliminate this information gain from our model in every case, we will try to minimize it instead. This leads to the maximum entropy method, which is guaranteed to produce a valid joint distribution from any pair of marginal distributions. Entropy is a concept from information theory and statistical physics that is used as a measure of disorder. For a bivariate distribution with probabilities p_{ij}, it is defined as

$$H = -\sum_{ij} p_{ij} \log p_{ij}, \qquad (2.8)$$

where the sum is over all states with non-zero probability. Equivalently, we may include zero-probability states if we employ the limit $\lim_{p \to 0} p \log p = 0$.

Maximizing the entropy H minimizes the amount of information we inject into the joint distribution over and above that which is already contained in the marginal distributions and the monotonicity constraint. One can prove that in the case of zero overlap between losses at t_1 and t_2, independence gives the joint distribution with maximum entropy. In the case of non-zero overlap, we construct the joint distribution via a numerical optimization, solving for the p_{ij} subject to the constraints of Eq. (1.3) and using the entropy of the joint distribution as the objective function.

The resulting joint distribution is much more dispersed than for the comonotonic method, and looking at the heat map in Fig. 3 one can immediately see that there is relatively little dependence between the losses at t_1 and t_2. For example, there is a significant probability of having a small initial loss at t_1 and yet a large loss at t_2.

2.4. *The Markov forward rate model*

Another approach to linking losses through time is the Markov forward loss transition rate model described by Schönbucher [3]. This is a fully skew-consistent dynamic approach in which the cumulative portfolio loss L is modelled directly rather than being constructing from the underlying single names. Transitions between different loss levels are controlled by a time-dependent generator matrix $A(t)$ whose off-diagonal elements $A_{ij}(t)$ for $i \neq j$ are forward loss transition rates giving the instantaneous rate at which probability mass flows from loss K_i to K_j at time t. The diagonal elements of the generator are determined by the requirement that the probability mass is conserved, which gives

$$-A_{ii}(t) = \sum_{\substack{k=0 \\ k \neq i}}^{N-1} A_{ik}(t) =: a_i(t). \qquad (2.9)$$

In this model, the portfolio loss distribution $p_j(t)$ evolves according to

$$\frac{dp_j}{dt} = -a_j(t)p_j + \sum_{\substack{k=0 \\ k \neq j}}^{N-1} A_{kj}(t)p_k, \qquad (2.10)$$

or in vector-matrix notation

$$\frac{d\mathbf{p}^T}{dt} = \mathbf{p}^T A, \qquad (2.11)$$

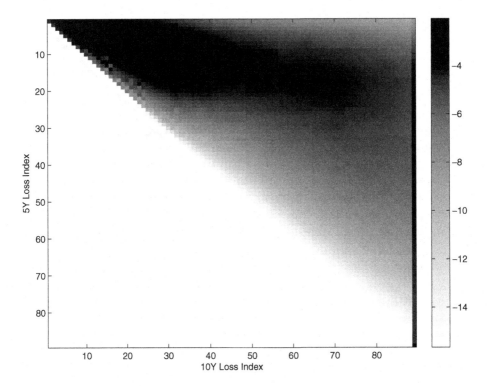

Fig. 3. Joint loss distribution for the maximum entropy method. Losses at 5 years and 10 years are represented along the vertical and horizontal axes respectively. Dark shades represent high probability. *Source*: Lehman Brothers.

where $\mathbf{p}^T$ is the row vector corresponding to $\mathbf{p}$. If the generator matrix A is independent of time, the solution to this equation is simply

$$\mathbf{p}^T(t_2) = \mathbf{p}^T(t_1) \exp\bigl(A(t_2 - t_1)\bigr), \qquad (2.12)$$

where the matrix exponential $\exp(X)$ is defined by the power series

$$\exp(X) := \sum_{n=0}^{\infty} \frac{X^n}{n!}. \qquad (2.13)$$

The matrix exponential can be calculated numerically using a variety of methods as reviewed in [2]. Of these, the most robust general-purpose approach appears to be the scaling and squaring algorithm and we have used a recent refinement of this technique described in [1].

In the loss transition rate framework, we can easily impose the monotonicity condition by using a generator matrix A which is upper-triangular with non-negative off-diagonal elements, i.e., $A_{ij}(t) = 0$ for $i > j$ and $A_{ij}(t) \geq 0$ for $i < j$. The diagonal elements are determined by the normalization condition given in Eq. (2.9). The simplest such generator matrix is one which only allows one-step transitions and has the bi-diagonal structure shown below

$$
A = \begin{pmatrix}
-a_0 & a_0 & 0 & \cdots & \cdots & 0 \\
0 & -a_1 & a_1 & 0 & \cdots & 0 \\
0 & 0 & -a_2 & a_2 & \cdots & 0 \\
\vdots & \vdots & \vdots & \ddots & \ddots & 0 \\
0 & \cdots & \cdots & \cdots & \cdots & a_{N-2} \\
0 & \cdots & \cdots & \cdots & \cdots & -a_{N-1}
\end{pmatrix}.
\tag{2.14}
$$

If we wish to conserve normalization in the evolution then we set $a_{N-1} = 0$, otherwise this element can take a general (positive) value representing the rate at which probability mass flows off the largest loss level K_{N-1} (we use this freedom in the calibration stage described below). The bi-diagonal generator matrix is completely determined by the vector $\mathbf{a}$ whose elements $a_i := -A_{ii}$ give the rate at which probability mass flows off one loss level and onto the next.

In the following, we restrict our attention to bi-diagonal generator matrices which are piecewise constant in time, so that Eq. (2.12) can be applied recursively to separate time intervals. To be precise, we suppose we have an ordered set of M times $t_0 < t_1 \cdots < t_{M-1}$, and a corresponding set of $M - 1$ constant generator matrices $\{A^i\}$ such that $A(t) = A^i$ for $t \in [t_i, t_{i+1})$. To evolve a given initial distribution $\mathbf{p}(T_0)$ between two times T_0 and T_1 satisfying $t_0 \le T_0 < T_1 \le t_{M-1}$ we proceed as follows: First, we find the minimum value of t_{i+1} such that $t_{i+1} > T_0$. Using the generator A^i we evolve $\mathbf{p}(T_0)$ to $\mathbf{p}(T)$ using Eq. (2.12), where the final time argument is $T = \min(t_{i+1}, T_1)$. If $T_1 > t_{i+1}$, we set $T_0 = t_{i+1}$ and repeat, otherwise we have $\mathbf{p}(T_1) = \mathbf{p}(T)$.

2.4.1. *Calibration*

The Markov loss rate model is calibrated by determining the set of generator matrices which enable it to match a given set of marginal loss distributions through time. The input to the calibration is a set of discrete loss distributions $p_i(t_j) := \mathrm{P}[L_{t_j} = K_i]$ defined on a set of M ordered dates $t_0 < t_1 \cdots < t_{M-1}$, and a set of N ordered strikes $K_0 < K_1 \cdots < K_{N-1}$. For each time interval we calibrate a constant generator matrix A^i, i.e. $A(t) = A^i$ for $t \in [t_i, t_{i+1})$. Since we only consider bi-diagonal generators this is equivalent to calibrating the piecewise constant vector $\mathbf{a}^i$ with elements $a_j^i = -A_{jj}^i$. Since we have exactly one free parameter for each strike and time the solution is uniquely determined.

The calibration proceeds via a boot-strapping procedure involving the recursive application of a root-finding algorithm. For each time interval $[t_i, t_{i+1})$, $i = 0, \ldots, M - 2$ we start from the probability distribution $\bar{\mathbf{p}}(t_i)$ generated by the model in the previous step (for $i = 0$, the vector $\bar{\mathbf{p}}(t_0)$ is an input which almost always corresponds to a deterministic distribution concentrated on the lowest strike).

We then calibrate the vector $\mathbf{a}^i$ by iterating over its elements a_j^i. Assuming that the values of a_j^i are known for $j < k$, we guess a value of a_k^i and evolve the loss distribution for strikes $i = 0, \ldots, k$ to obtain $\bar{\mathbf{p}}(t_{i+1})$ using Eq. (2.12). We then vary the value of a_k^i until we match the input loss probability $p_k(t_{i+1})$. This is straightforward because $p_k(t_{i+1})$ is a monotonically decreasing function of a_k^i so there is either a unique solution (if the input is arbitrage-free) or none (if the input distribution contains arbitrage). In the latter case, we replace a_k^i with a cap or a floor as appropriate (in this manner, the calibration can actually 'repair' any arbitrage in the input loss distributions). At each stage of the recursion we only evolve the portion of the loss distribution required, so we deal with a subset of the full loss support consisting of loss levels K_i for $i = 0, \ldots, k$. During the calibration of element a_k^i, the generator matrix is therefore of size $(k + 1) \times (k + 1)$ and it grows by one row and one column on each iteration.

Formally, the transition rates are defined on the non-negative real axis, $a_j^i \in R^+$, so a direct root-search on these quantities is not practical numerically. We overcome this problem by mapping the rates to the interval $[0, 1]$ and defining the root search in terms of the mapped rates, which is then a well-posed numerical problem. We also combat numerical instabilities arising from extreme transition rates by introducing a cap and a floor on their permitted values. The chosen limits easily encompass the range of transition rates and implied spreads we might reasonably expect to see in the market so this restriction does not have a noticeable effect on the predictions of the model. In practice, we only observe the transition rates hitting these bounds when the loss probability is negligibly small, in which case the transition rates are rather ill-defined and their precise value is not of great importance.

2.4.2. *Calculation of joint distributions*

Once we have calibrated the generator matrices, we can easily calculate the joint distribution of the losses at two time horizons as defined in Eq. (1.2). For each loss level K_i at time t_1 we evolve the deterministic initial condition $p_j(t) = \delta_{ij}$ to time t_2. This gives the loss distribution at t_2 conditional on the loss at t_1. Scaling by the known marginal distribution at t_1 gives the required joint distribution. An example of a joint loss distribution obtained from the Markov loss rate model is shown in Fig. 4.

3. Comparison of the Different Methods

Given the range of loss-linking methods at our disposal, we would like to describe the dependence introduced by the different methods in a way that is suitable for comparison. We use the same bespoke portfolio described in Sec. 1.

3.1. *Correlation between future losses*

First, the correlation between losses at t_1 and t_2 provides a broad and intuitively accessible measure of dependence. We have calculated both linear correlation and

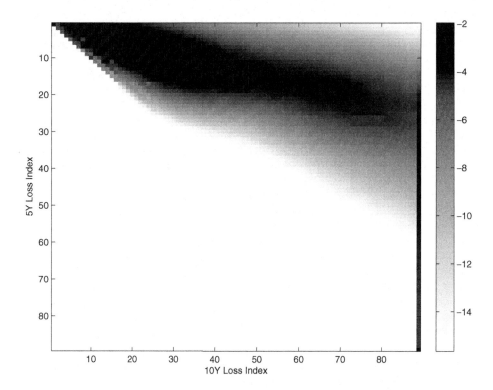

Fig. 4. Joint loss distribution for the Markov forward rate method. Losses at 5 years and 10 years are represented along the vertical and horizontal axes respectively. Dark shades represent high probability. *Source*: Lehman Brothers.

Spearman's rank correlation for losses at 5Y and the loss increment between 5Y and subsequent maturities, using

$$q_n = \sum_{k=0}^{N-n-1} p_{k,k+n}, \tag{3.1}$$

to convert the joint distribution to the distribution of loss increments (cf. Eq. (2.5) for the definition of q_n), where K_{N-1} is the maximum portfolio loss.

Of the three methods we tested, the maximum entropy method gives the lowest correlation, although the value of 38.6% for a final maturity at 10Y is still significantly above zero. This is a reflection of the fact that the method of independent increments is not possible for the 5Y to 10Y period. If it were possible to have independent increments, the corresponding joint distribution would coincide with the solution obtained from maximum entropy and it would obviously have a correlation of 0%.

The comonotonic method results in a very high positive correlation, as expected from the way it is constructed. The Markov forward rate model gives correlations in between the other two methods but is slightly closer to maximum entropy.

Table 1. Correlation of the loss at 20 June 2011 and the loss increment to later dates.

	Linear Correlation			Spearman's Rank Correlation		
	20-Jun-12	20-Jun-14	20-Jun-16	20-Jun-12	20-Jun-14	20-Jun-16
Comonotonic	74.3%	71.0%	67.9%	84.4%	95.9%	97.6%
Markov	55.0%	57.0%	52.0%	32.8%	45.9%	55.4%
Max Entropy	40.1%	45.9%	40.9%	18.7%	29.5%	38.6%

Source: Lehman Brothers.

Table 2. Entropy of joint distributions of the loss at 20 June 2011 and later.

Date	Marginals	Comonotonic	Markov	Max Entropy
20-Jun-11	2.848	2.848	2.848	2.848
20-Jun-12	3.084	3.510	4.799	4.971
20-Jun-13	3.241	3.657	5.210	5.406
20-Jun-14	3.418	3.759	5.521	5.721
20-Jun-15	3.598	3.919	5.767	5.967
20-Jun-16	3.739	3.965	5.950	6.148

Source: Lehman Brothers.

As a function of the second maturity, rank correlation is increasing for all three methods. However, there is no clear trend in terms of linear correlation. Since the loss distributions are highly non-Gaussian, rank correlation is the preferred correlation measure because it is purely a function of the copula joining the losses and is independent of the marginals.

3.2. *Entropy of the joint distribution*

In Table 2 we list the entropy for all three methods, and as a reference the entropy of the marginal distributions is also included. In terms of the entropy measure of dependence, the order of the three methods is reversed. Obviously, maximum entropy gives the highest entropy. Of the other two methods, the Markov forward rate model has higher entropy than the comonotonic method. The latter has a low entropy because it is highly structured in the sense that there is very little 'randomness' in the joint distribution: knowing the loss at t_1 determines the loss at t_2 and vice versa, up to discretisation effects. While the Markov forward rate model again falls in between the other two methods, it is much closer to maximum entropy according to this measure.

For the three loss-linking methods we discuss in this article, correlation and entropy form an inverse relation. However, this is not always the case. Entropy should therefore not be seen as an inverse proxy for correlation. It is a measure of disorder and accordingly it is possible to construct highly structured joint distributions which have low correlation as well as low entropy. Note also that distributions with

Table 3. Properties of the loss increment distributions for the period 5Y to 10Y.

	Comonotonic	Markov	Max Entropy
Expected Loss Increment	4.15%	4.15%	4.15%
Standard Deviation	2.32%	2.51%	2.66%

Source: Lehman Brothers.

negative correlation exist, at least for some marginal distributions, whereas entropy is always positive.

As a function of the second maturity, the entropy is strictly increasing for all three methods. This is because the marginal distribution at 5 years contains progressively less information about the loss distribution at the later date.

3.3. *Loss increment distributions*

It is instructive to look at the distribution of loss increments more closely. The expected loss increment is independent of how we link losses through time as it depends only on the marginal distributions at t_1 and t_2,

$$\mathrm{E}[\Delta L(t_1, t_2)] = \mathrm{E}[L_{t_2}] - \mathrm{E}[L_{t_1}]. \tag{3.2}$$

In our case, the expected loss increment for the 5Y to 10Y period is 4.15% of the portfolio notional. While the expected loss is not model-dependent, the standard deviation of the loss increment distribution does depend on how we link losses through time. The results in Table 3 show that the distribution is widest for the maximum entropy method and narrowest for the comonotonic method. This has implications for pricing forward-starting CDOs, as we will see in the next section.

The full loss increment distributions for the comonotonic and maximum entropy methods are shown in Fig 5.

4. Forward Starting CDOs

Forward starting CDOs (referred to as FDOs in the following) provide a simple application of our modelling framework. In an FDO, the investor sells protection on a CDO tranche starting at a future date t_1 (the protection start date) and extending to a final maturity T. However, FDOs come in two flavours: If the contract specifies that losses occurring before t_1 lead to a reduction of the tranche subordination, we can replicate the FDO by selling protection on a CDO tranche with maturity T and buying protection on a CDO tranche with maturity t_1. Hence, the price is uniquely determined by the marginal distributions. Here, we are concerned with FDOs of the second kind for which the subordination of the tranche is increased to offset any losses between the value date and the protection start date. The pricing

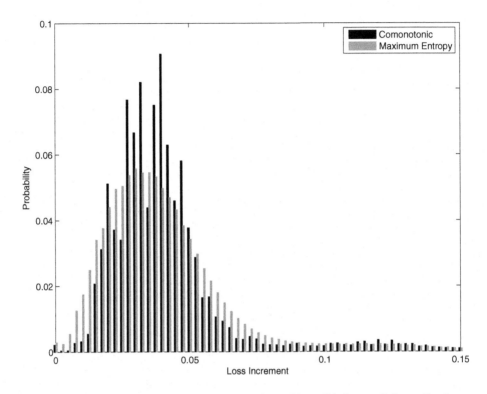

Fig. 5. Loss increment distribution for the period from 5Y to 10Y. Source: Lehman Brothers.

of the latter type of FDO depends only on the distribution of the loss increments $L(t_1, t_2)$ with $t_2 \in [t_1, T]$. This can be seen as follows:

At any time t, the tranche loss is

$$L_t^{Tr} = \left(L_t - K_A(t)\right)^+ - \left(L_t - K_D(t)\right)^+, \qquad (4.1)$$

where $K_A(t)$ and $K_D(t)$ are the time-dependent attachment and detachment points of the tranche in question. For a refreshing tranche these are given by

$$K_A(t) = K_A + L_{t_1}, \quad K_D(t) = K_D + L_{t_1}, \qquad (4.2)$$

for $t \geq t_1$ and constant K_A and K_D. It follows that the tranche loss at any time $t_2 \geq t_1$ can be written as

$$L_{t_2}^{Tr} = \left(\Delta L(t_1, t_2) - K_A\right)^+ - \left(\Delta L(t_1, t_2) - K_D\right)^+, \qquad (4.3)$$

which depends only on the loss increment. As the expected loss increment is model-independent, the main determinant of value in the pricing of FDO tranches is the width of the loss increment distribution. As the maximum entropy method, with a standard deviation of 2.66%, gives the widest distribution it implies high risk for the upper end of the capital structure and low risk for the lower end relative to the other two methods. Similarly, the comonotonic method, which leads to a narrow loss increment distribution, implies relatively high risk at the low end and low risk at the upper end of the capital structure.

The different widths of the loss increment distribution are reflected in the breakeven spreads that we calculate for a series of thin FDO tranches with a protection start date at 5Y and a final maturity of 10Y. The results are shown in Table 4 where we have set the highest breakeven spread for a given tranche in bold and the lowest spread in italics. Of particular interest is the inflection point as the order of the breakeven spreads reverses when going from the $3 - 4\%$ to the $4 - 5\%$ tranche. This coincides with the expected loss increment at 4.15%.

As a simple illustration of the effects of a narrow loss increment distribution, consider the case where the loss increment is deterministic. In this limiting case, a thin tranche below the expected loss increment is certain to be wiped out while a thin tranche above this point is certain to survive. A narrow loss increment distribution therefore implies high risk for junior tranches and low risk for senior tranches, consistent with our results above.

As the comonotonic method is based on maximum dependence, we regard the breakeven spread it predicts as providing a valid bound in practical circumstances. Similarly, the maximum entropy method, giving in some sense a minimum dependency, could be seen as an approximate bound in the other direction. However, two cautionary notes are in order. First, the maximum entropy method gives a much weaker bound than the comonotonic method, as can be seen from the following argument: By its nature, the comonotonic method introduces high positive correlation between losses at different maturities, while maximum entropy leads to a small positive correlation. However, it is in fact possible to construct joint distributions with negative correlation. These would typically give a wider loss increment distribution than maximum entropy as the probability for very large loss increments increases. The maximum entropy method should therefore be viewed as producing the joint distribution that is closest to independence while respecting the no-arbitrage constraints.

Table 4. Breakeven spreads for a 5-into-5 forward-starting CDO.

Tranche	Comonotonic	Markov	Max Entropy
0–1%	**1500**	1443	*1413*
1–2%	**1113**	1059	*1034*
2–3%	**790**	749	*727*
3–4%	**501**	489	*477*
4–5%	*250*	282	**288**
5–6%	*118*	153	**166**
6–7%	*75*	92	**102**
7–8%	*59*	66	**71**
8–9%	*50*	53	**56**
9–10%	*42*	44	**45**

Source: Lehman Brothers.

Second, we also note that even the spreads predicted by the comonotonic method do not represent a strict bound as it is likely that we can construct joint distributions which concentrate incremental losses more heavily in any particular region of the capital structure. However, we believe that these joint distributions are not likely to represent the loss distribution in realistic situations.

5. Conclusions

We have discussed a range of ways to value exotic correlation products where the value depends on the joint distribution of the portfolio loss at two or more time horizons. The Markov forward rate model is a skew-consistent arbitrage-free model which implies a joint distribution once calibrated to vanilla CDO prices. The comonotonic and maximum entropy methods impose a dependency structure exogenously. While the comonotonic method gives the maximum dependence between future losses, the maximum entropy method is designed to give the minimum dependence.

As an application of our modelling framework, we have priced forward-starting CDO tranches using the different loss-linking methods. The comonotonic and maximum entropy methods can be regarded as providing reference points for the breakeven spreads. While the comonotonic method provides an upper bound for equity tranches and a lower bound for senior tranches in realistic cases, the maximum entropy method gives low spreads for equity tranches and high spreads for senior tranches. For a thin tranche, the crossover occurs when the strike equals the expected loss increment over the life of the trade.

References

[1] N. J. Higham, The scaling and squaring method for the matrix exponential revisited, *Siam J. Matrix Anal. Appl.* **26** (2005) 1179–1193.
[2] C. Moler and C. Van Loan, Nineteen dubious ways to compute the exponential of a matrix, twenty-five years later, *Siam Review* **45** (2003) 3–49.
[3] P. J. Schönbucher, Portfolio losses and the term structure of loss transition rates: A new methodology for the pricing of portfolio credit derivatives, Working paper, Department of Mathematics, ETH Zürich (2005).

ON THE TERM STRUCTURE OF LOSS DISTRIBUTIONS: A FORWARD MODEL APPROACH

JAKOB SIDENIUS

J P MORGAN Securities, 125 London Wall
London EC2Y 5AJ, United Kingdom
jakob.m.sidenius@jpmorgan.com

We define forward copula models and introduce the concept of "chaining" such models. We discuss the use of these concepts in the calibration to the term structure of tranche quotes.

Keywords: CDO pricing; copula model; forward copula; copula chain; CDO term structure; forward base correlations.

1. Introduction

In recent years the tranches of the standard index pools of CDS's as ITraxx and CDX have become increasingly liquid and have begun trading at several standard maturities. Although the five year maturity remains the most liquid both shorter and longer maturities trade in sufficient volumes to be able to serve as market references. This allows on the one hand dealers to risk manage their seasoned books and on the other hand investors to take views on the relative pricing of different maturities.

Another development has been the emergence of new products sensitive not only to the term structure, but also to the dynamics of losses and loss distributions. A few examples of such products are forward starting tranches, tranche options and leveraged tranches. In the modeling of this dynamics such as, eg, in [9, 10], one takes — implicitly or explicitly — as the starting point the initial term structure of loss distributions.

One problem with modeling this term structure is that in the market only a handful of tranches are quoted and trade liquidly for each of only a few maturities. Thus the pricing of tranches of other seniorities and/or maturities has to rely on some form of interpolation. Since tranches are quoted on par spreads, one cannot interpolate directly in the market quotes and since the conversion from par spread to (leg) value requires a pricing model, one cannot interpolate in values either.

For a single maturity liquid tranche prices can be matched by base-correlations [8], but again there is no interpolation rule which can guarantee consistency (absence of arbitrage).

A number of factor copula models have been proposed attempting to match simultaneously all liquid tranches with a common maturity, see for example [2, 5–7, 11]. Since these models are consistent by construction, any tranche can be priced consistently with such a model. However, unless the maturity is close to the common maturity, the fit to the market is unsatisfactory. Furthermore, attempts to repair this situation by letting model parameters become maturity dependent are easily seen to lead to inconsistencies.

In this paper we present a way of combining the calibrations to several maturities in a single model. To do this we use a new formalism for building models as chains of models each of which describes the loss distribution over a forward interval (between tranche maturities). All the "link" models are assumed to be factor copula models, i.e., model default times are independent when conditioned on the model factors. The chain is again a factor copula model and as such supports the use of standard techniques [4] for pricing and sensitivity computations.

The formalism supports more general chain constructions than required for matching the term structure of tranche prices. This can be used to construct chain models for which one can control the decoupling of losses in different forward periods (see also very recent paper [1] for discussions of such models).

In the next section we define forward copulas and proceed in Sec. 3 to idea of chaining such models together to form a new factor copula model. We present some special cases of this idea in Sec. 4 where we also give some important applications. Section 5 contains our concluding remarks. An appendix contains some useful observations about factor copula models.

2. Forward Copula Models

We begin by considering in complete generality a factor model of the default times τ_i in a given portfolio of N assets labeled $i = 1, \ldots, N$. Let there be given

- a factor Z with distribution F_Z over some domain D_Z (possible multidimensional);
- a set of continuous functions

$$g_i(t, z) : [0, \infty[\times D_Z \to]0, 1], \quad i = 1, \ldots, N,$$

with the properties

$$\forall z \in D_Z : g_i(0, z) = 1;$$
$$\forall z \in D_Z : \lim_{t \to \infty} g_i(t, z) = 0;$$

Then we can interpret the g_i's as (factor) conditional survival curves, i.e.,

$$g_i(t, z) = P(\tau_i > t|z),$$

such that

$$P(\tau_i > t) = \int_{D_Z} g_i(t, z) dF_Z(z). \tag{2.1}$$

We shall furthermore require conditional independence of the default times, i.e.,

$$P(\tau_i > t, \tau_j > s | z) = g_i(t, z) g_j(s, z).$$

With this definition a factor model implies a definite joint distribution of default times and guarantees the consistency of loss distributions to all horizons. The default time distribution can be straightforwardly sampled by Monte Carlo and loss distributions can be computed by convolution techniques [4].

The above definition (see also [3] may be unfamiliar, but for the purposes of this paper it is more convenient than the usual definition in terms of default trigger variables. The latter is often useful for implementations, however, because it gives a concrete parameterization which can be used in calibration — see Appendix Appendix A for details.

To define a forward starting model we exploit the usual analogy between (conditional) survival probabilities and discount factors. Thus we can define a forward model with forward start at time $T > 0$ by the conditional survival curves

$$g_i^{(T)}(t, z) := g_i(t, z) / g_i(T, z) : [T, \infty[\times D_Z \to]0, 1], \quad i = 1, \ldots, N. \tag{2.2}$$

We now have

$$g_i^{(T)}(T, z) = 1;$$
$$\lim_{t \to \infty} g_i^{(T)}(t, z) = 0;$$
$$g_i^{(T)}(t, z) = P(\tau_i > t | \tau_i > T, z), \quad t \geq T. \tag{2.3}$$

Motivated by this observation we shall use the properties (2.3) of the conditional survival curves to define a forward starting copula factor model. Our aim in the next section is to construct a "chain" of forward models covering disjoint forward intervals such that the chain model covers all positive times and such that we have (2.3) for all forward models. Thus we want the conditional survival curves of each forward model to give the survival probability conditioned on the factor and on survival to forward start.

3. Chaining

Consider first the case of just two models: one spot starting, the other forward starting at time $T > 0$. Let the models be given by quadruplets $\left(Z^{(k)}, D_{Z^{(k)}}, F_{Z^{(k)}}, g_i^{(k)} \right)$ with superscript (1) ((2)) referring to the spot (forward) starting model. The conditional survival curves g_i of the "chain" model will depend on both factor values $z^{(1)}$ and $z^{(2)}$, but for times before T we have simply

$$g_i(t, z^{(1)}, z^{(2)}) := P\left(\tau_i > t | z^{(1)}, z^{(2)} \right) = g_i^{(1)}(t, z^{(1)}), \quad t < T,$$

with no dependence on $z^{(2)}$. However, for $t > T$ we have

$$
\begin{aligned}
g_i(t, z^{(1)}, z^{(2)}) &:= P\big(\tau_i > t | z^{(1)}, z^{(2)}\big) \\
&= P\big(\tau_i > t | \tau_i > T, z^{(1)}, z^{(2)}\big) P\big(\tau_i > T | z^{(1)}, z^{(2)}\big) \\
&= P\big(\tau_i > t | \tau_i > T, z^{(1)}, z^{(2)}\big) g_i^{(1)}(t, z^{(1)})
\end{aligned}
$$

and so we must make the identification

$$
g_i^{(2)}(t, z^{(2)}) = P\big(\tau_i > t | \tau_i > T, z^{(1)}, z^{(2)}\big). \tag{3.1}
$$

This last equation looks strange since the right hand side depends on $z^{(1)}$ whereas the left hand side does not. The only way out is to require that the factor structure of the forward model "contain" the factor structure of the spot model. More precisely, we require that for some suitable $\tilde{Z}^{(2)}$

$$
\begin{aligned}
Z^{(2)} &= (Z^{(1)}, \tilde{Z}^{(2)}); \\
D_{Z^{(2)}} &= D_{Z^{(2)}} \times D_{\tilde{Z}^{(2)}};
\end{aligned} \tag{3.2}
$$

and that the marginal distribution of $Z^{(1)}$ in $F_{Z^{(2)}}$ is $F_{Z^{(1)}}$. When this is the case (3.1) makes sense because we can rewrite is as

$$
g_i^{(2)}(t, z^{(1)}, \tilde{z}^{(2)}) = P\big(\tau_i > t | \tau_i > T, z^{(1)}, \tilde{z}^{(2)}\big) \tag{3.3}
$$

which allows to define the chain model as the factor copula model having factor $Z = Z^{(2)}$ and conditional survival curves

$$
g_i(t, z) := \begin{cases} g_i^{(1)}(t, z^{(1)}), & t \leq T \\ g_i^{(2)}(t, z^{(2)}) g_i^{(1)}(T, z^{(1)}), & t > T \end{cases} \tag{3.4}
$$

The chaining mechanism just developed is very general and the requirements made on "link" models are — perhaps contrary to perceptions — very weak since the factor structure of a link can always be trivially extended to satisfy them.

More specifically, if we are given two triplets (Z, D_Z, F_Z) and (Y, D_Y, F_Y) we can extend the second factor structure such that it includes the first (or *vice versa*) in the sense required for chaining. The important point is, of course, that we can do so without in any way changing the corresponding link model. To see this define the trivially extended factor structure by

$$
(\hat{Y}, D_{\hat{Y}}, F_{\hat{Y}}) = ((Y, Z), D_Y \times D_Z, F_Y \times F_Z)
$$

and define the conditional survival curves for the trivially extended model as

$$
\hat{g}_i(t, \hat{y}) = g_i(t, y),
$$

i.e., independent of Z.

The above remark shows that we can chain arbitrary models without making any assumption about their factor structures. We can think of this way of chaining as just taking a "straight" product of the link models where the factor structures of different links are independent. However, the general chaining mechanism allows to "twist" the product if the link factor structures support it. In the next section we shall look at some examples of this.

The chain model produced by the fusing of the two links is again a factor copula model as defined in Sec. 2. As such it may, in turn, be considered as a link to be fused with a third model and so on. Thus there is no absolute distinction between chain models and link models and there is no need for — nor indeed any room for — an independent definition of multi-link chains; the result of fusing several links follows from the procedure described above.

Let us anyway briefly consider the result of fusing a chain of K links. We consider a set of maturities $0 = T_0 < T_1 < \cdots < T_K$ and, for each interval $[T_{k-1}, T_k]$, $k = 1, \ldots, K$, a forward model for which the factor $Z^{(k)} = (Z^{(k-1)}, \tilde{Z}^{(k)})$ takes values in $D_{Z^{(k)}}$ with distribution $F_{Z^{(k)}}$ such that the marginal distribution for $Z^{(k-1)}$ is given by $F_{Z^{(k-1)}}$. The conditional survival curves of the kth model are given by functions $g_i^{(k)}(t, Z^{(k)})$. The factor structure of the chain model out to and including the kth forward model is given by the triplet $(Z^{(k)}, D_{Z^{(k)}}, F_{Z^{(k)}})$ and the conditional survival curves by

$$g_i(t, z^l) = g_i^{(l)}(t, z^{(l)}) \prod_{j=1}^{l-1} g_i^{(j)}(T_j, z^{(j)}), \quad T_{l-1} < t \leq T_l,$$

and one verifies by inspection that indeed

$$g_i(t, z^l) = P(\tau_i > t | z^l), \quad T_{l-1} < t \leq T_l.$$

Note that for $T_{k-1} < t \leq T_k$ the dependence of $g_i(t, z^l)$ on $Z^{(K)}/Z^{(l)}$ is trivial.

We conclude this section by discussing the important issue of the calibration of chain factor models to single-name CDS curves, i.e., to marginal default time distributions. This can be done in a straightforward "bootstrap" procedure. We must have [see (2.1)]

$$P(\tau_i > t) = \int_{D_{Z^{(k)}}} g_i^{(k)}(t, z^{(k)}) \prod_{j=1}^{k-1} g_i^{(j)}(T_j, z^{(j)}) dF_{Z^{(k)}}(z^{(k)}), \quad T_{k-1} < t \leq T_k,$$

so for $k = 1$ we have to search, for each asset, for (parameters of) $g_i^{(1)}$ such that

$$P(\tau_i > t) = \int_{D_{Z^{(1)}}} g_i^{(1)}(t, z^{(1)}) dF_{Z^{(1)}}(z^{(1)}), \quad t \leq T_k. \tag{3.5}$$

Once the $g_i^{(1)}$'s have been thus determined we proceed to determining the $g_i^{(2)}$'s by requiring, for each i,

$$P(\tau_i > t) = \int_{D_{Z^{(2)}}} g_i^{(2)}(t, z^{(2)}) g_i^{(1)}(T_1, z^{(1)}) dF_{Z^{(2)}}(z^{(2)}), \quad T_1 < t \leq T_2, \tag{3.6}$$

and so on.

The only non-trivial point is whether it is guaranteed that we can find $g_i^{(2)}$'s etc such that (3.6) is satisfied. Note that the issue is not so much whether we can satisfy the requirement for all t, but rather whether we can do it for all t such that the conditional survival curves are decreasing for all z. We shall demonstrate that this is so in the case most interesting in practice, where $g_i^{(2)}$ are defined as the forward survival curves of a factor model defined by a trigger variable specification. In this case we have the explicit form [from (A.3) and (2.2)]

$$
g_i^{(2)}(t, z^{(2)}) = \frac{1 - F_i^{(2)}\left(h_i^{(2)}(t) - f_i^{(2)}(z^{(2)})\right)}{1 - F_i^{(2)}\left(h_i^{(2)}(T_1) - f_i^{(2)}(z^{(2)})\right)},
\tag{3.7}
$$

where $f_i^{(2)}$ is some function introduced to allow RFL type specifications. Here we wish to choose a continuous $h_i^{(2)}(t)$ defined on $[T_1, T_2]$ such that (3.6) holds. First, note that the value of $h_i^{(2)}(T_1)$ is not determined by (3.6). Its value is, in fact, arbitrary and will not affect the resulting model.[1] Next, from the explicit form (3.7) we see that for any fixed t and z, $g_i^{(2)}$ is decreasing in $h_i^{(2)}(t)$ and therefore so is the integral on the right hand side of (3.6). Furthermore, since $g_i^{(2)}$ depends on t only through $h_i^{(2)}(t)$ so does the integral. Now the left hand side is given and the right hand side is decreasing in $h_i^{(2)}(t)$. Possible given values for the left hand side lie in the interval $]0, P(\tau_i > T_1)]$ and this is covered by the right hand side by letting $h_i^{(2)}(t)$ vary over $[h_i(T_1), \infty[$. Thus a solution does exist and since the given values of $P(\tau_i > t)$ are decreasing in t, the solutions for $h_i^{(2)}(t)$ must be increasing in t. Thus we see that the $g_i^{(2)}$ which solves (3.6) is indeed decreasing in t for any z.

We note that it will be very hard to preserve any tractability in the chain model beyond the first period. This means that all factor integrals will have to be evaluated numerically. This makes the tendency of general chain models to "grow more factors" with horizon highly problematic for their practical usefulness. In fact, unless we restrict ourselves to some special case (see next section) we cannot expect chain models with more than two periods to escape the "curse of dimensionality" and be manageable in practice.

4. Special Cases and Examples

Here we shall give details of some special cases and examples of the general chain construction. For ease of exposition we restrict ourselves to the case of just two links in the chain, but the extension to general chains is obvious. We use the notation introduced above: the two models are given by quadruplets $\left(Z^{(k)}, D_{Z^{(k)}}, F_{Z^{(k)}}, g_i^{(k)}\right)$ with superscript (1) ((2.2)) referring to the spot (forward) starting model. In all cases the forward model is defined from a spot model as in (2.2).

---

[1] Although, of course, it will affect $h_i^{(2)}(t)$ for $t > T_1$.

4.1. *Cylindrical chain*

Suppose that (up to isomorphism) $D_{Z^{(2)}} = D_{Z^{(1)}}$ and $F_{Z^{(2)}} = F_{Z^{(1)}}$. Then we can simply identify the factors, i.e., define the common factor $Z = Z^{(1)} = Z^{(2)}$, and this will not in any way affect either of the two models when viewed in isolation. When forming the chain it simply means that $\tilde{Z}^{(2)}$ in (3.2) should be the trivial random variable and that the factor structure of the chain model is given by the common factor.[2] We call this type of chain "cylindrical" because the factor structure is the same for all horizons.

In practice the link models will be given by trigger variable specifications and marginal default time distributions matched by choosing suitable threshold functions $h_i^{(k)}$ on $[T_{k-1}, T_k]$. As noted above, $h_i^{(k)}(T_{k-1})$ is arbitrary, and we are free to adopt the convention

$$h_i^{(k)}(T_{k-1}) = h_i^{(k-1)}(T_{k-1}), \tag{4.1}$$

where the right hand side *is* determined by (3.5). The reason for the choice (4.1) is that now, if the $k-1$st and kth models have identical copula structures, the chain is manifestly equivalent to a single copula model. In this sense we can say that, at least for copulas with continuous parameter ranges, e.g., RFL, the cylindrical chain is a continuous deformation away from the single-period model. We point out that when the two consecutive link models do not have the same copula parameters, then the chain in general cannot be defined using the trigger variable set-up of the link models — in this sense the chain is "outside the family" of the link models. This is, of course, the whole point of the chaining formalism.

4.1.1. *Term structure of base correlations*

Base correlations [8] are defined in terms of the one-factor Gaussian copula model as the implied flat correlations for generalized equity tranches with a common maturity, T. Thus to a given detachment level d corresponds a certain Gaussian copula model defined such that the tranche with maturity T and detachment level d is correctly priced. By replacing the Gaussian copula model by a cylindrical chain of Gaussian copula models we can extend this construction to more than one maturity.

Here we consider just two maturities $T_1 < T_2$ and we wish to construct a chain model which matches the given prices of two tranches with maturities T_1 and T_2 and a common detachment level d. This can be done by a straightforward "bootstrapping" of a cylindrical chain of two Gaussian copula models with flat correlations ρ_1 and ρ_2, respectively. First set ρ_1 to the standard base correlation for detachment level d and maturity T_1. Next set ρ_2 such that the price of the tranche with detachment d and maturity T_2 is matched. Note that both ρ_1 and ρ_2 can be found by straightforward root search.

---

[2]Note that in the sense of Sec. 3 the cylindrical chain is "maximally twisted".

Specifically, we have then constructed a model with Gaussian factor Z and conditional survival curves [this is directly from (2.2) and (3.4)]

$$
g_i(t, z) := \begin{cases} 1 - \Phi\left(\frac{h_i(t) - \sqrt{\rho_1}z}{\sqrt{1 - \rho_1}}\right), & t \leq T_1 \\ \frac{1 - \Phi\left(\frac{h_i(t) - \sqrt{\rho_2}z}{\sqrt{1 - \rho_2}}\right)}{1 - \Phi\left(\frac{h_i(T_1) - \sqrt{\rho_2}z}{\sqrt{1 - \rho_2}}\right)} \left(1 - \Phi\left(\frac{h_i(T_1) - \sqrt{\rho_1}z}{\sqrt{1 - \rho_1}}\right)\right), & T_1 < t \leq T_2 \end{cases}, \quad (4.2)
$$

where $h_i(t)$ is a function calibrated such that unconditional survival probabilities are matched. Note that as $\rho_1 \to \rho_2$ the chain becomes equivalent to a single Gaussian copula model. As noted above this is a general feature of cylindrical models.

Because of its definition as the correlation parameter of a forward starting model ρ_2 is in a very real sense a forward base correlation and the chain model can be used to consistently price tranches with detachment level d for any maturity. Note however, that the introduction of forward base correlations offers no clues to the pricing of tranches with other detachment levels. Loosely speaking, we can say that it solves only the problem of interpolating in maturity and not the problem of interpolating in detachment level. Of course, this can still be regarded as progress compared to the case of all-spot-starting base correlations for which one is reduced to using heuristic interpolation rules attended by laborious checks for inconsistencies (arbitrage).

4.1.2. Term structure of loss distributions

Suppose we have a factor copula model which is capable of matching to good approximation the prices of a set of tranches with common maturity T. Then we can set up a cylindrical chain of such models to try and match a number of tranches at each of a number of maturities. This model then allows us to price in a consistent way tranches of arbitrary maturity and detachment level. We now briefly discuss how such a model is calibrated in "bootstrap" procedure with a parameter search at each step. Note that, once the model has been calibrated, pricing and sensitivity computations proceed as usual for a factor copula model.

To keep the notation simple we denote the parameters of the copula of the kth model by $\alpha^{(k)}$, $k = 1, 2$. We assume that tranche quotes $Q^{(k)}$ are given for each maturity T_k. We begin at maturity T_1 by searching for $\alpha^{(1)}$ such that $Q^{(1)}$ are matched to the desired accuracy. Note that, as usual, for each trial value of $\alpha^{(1)}$ we need to calibrate the $g_i^{(1)}$'s to single-name survival curves over $[0, T_1]$.

Next we proceed to searching for values of $\alpha^{(2)}$ such that $Q^{(2)}$ are matched to the desired accuracy. Again, for each trial value of $\alpha^{(2)}$ we need to calibrate the $g_i^{(2)}$'s to single-name survival curves over $[T_1, T_2]$. But the crucial observation is that the calibration in the second step in no way disturbs the match obtained in the first step. In fact, once the spot model has been calibrated, it is efficient to cache $g_i(T_1, z)$ (for all z-nodes in the quadrature) as well as judiciously chosen time T loss expectations (depending on the considered tranches) since these will not change during the forward model calibration.

Clearly, the choice of an appropriate model for the links of this chain is crucial for the practical success of such a term structure model. In particular, the model should not be too constrained by the choice of factor distribution. Since the RFL specification in a sense contains all other specifications (see Appendix Appendix A), using an RFL type link model provides a great deal of useful flexibility.

For realistic computational loads the link model should have only one or two factors and the calibration to tranche quotes must not be too cumbersome. Note that the term structure model is designed to fit the observed market quotes with a parsimonious parameterization and to provide consistent implies loss distributions for all horizons, but that it does not allow any control over the degree of decoupling between losses in different periods. In the next section we shall see how such control may be retained, albeit at the cost of a considerable increase in computational complexity.

4.2. *Conical chain*

In the case where $\tilde{Z}^{(2)}$ in (3.2) is non-trivial, the factor structure of the forward model strictly includes that of the spot model and so the factor structure "grows" with horizon, creating a (half-)"cone" with the spot model at its apex. In a conical chain each new period adds "its own" factor which does not affect previous periods and this allows some explicit control over the decoupling of losses between periods.

We remark that since one or both of the models in the conical chain may themselves be chains, it is possible to mix cylindrical and conical subchains in the same model. This is useful if one needs to match tranche quotes at a number of horizons, but do not need to decouple all periods.

To illustrate the difference to the case of cylindrical chains we give three examples, all involving Gaussian copulas, beginning with the simplest case of the "straight" product of two Gaussian copula models. These three examples should be compared to the cylindrical Gaussian chain in Sec. 4.1.1. The case of conical chains with general links is straightforward in principle; in practice, as usual, much depends on the details of model specifications.

4.2.1. *Straight Gaussian chain*

Here the factor of the chain model is

$$Z = (Z^{(1)}, \tilde{Z}^{(2)}), \tag{4.3}$$

with the components being iid standard Gaussian. The conditional survival curves are given by [see (3.4)]

$$g_i(t,z) := \begin{cases} 1 - \Phi\left(\frac{h_i(t) - \sqrt{\rho_1}z^{(1)}}{\sqrt{1-\rho_1}}\right), & t \leq T_1 \\ \frac{1-\Phi\left(\frac{h_i(t) - \sqrt{\rho_2}\tilde{z}^{(2)}}{\sqrt{1-\rho_2}}\right)}{1-\Phi\left(\frac{h_i(T_1) - \sqrt{\rho_2}\tilde{z}^{(2)}}{\sqrt{1-\rho_2}}\right)}\left(1 - \Phi\left(\frac{h_i(T_1) - \sqrt{\rho_1}z^{(1)}}{\sqrt{1-\rho_1}}\right)\right), & T_1 < t \leq T_2 \end{cases} \tag{4.4}$$

We note that strong resemblance to (4.2): the "only" difference is that in (4.4) each period has "its own" factor. This difference, however, means that while factor integrals of the cylindrical Gaussian chain are always one-dimensional, the dimensionality grows linearly in the number of periods for conical chains. This feature makes conical chains with more than one or two links very costly in practical use.[3]

4.2.2. *Factor-coupled Gaussian chain*

If we take the factor structure to be a general bivariate Gaussian, i.e., with correlation $\neq 0$, introduce some copuling between the factors. We can control this coupling via the explicit correlation parameters, and in this sense this model sits between the straight chain just discussed and the cylindrical chain in Sec. 4.1.1; these two extremes correspond to factor correlation 0 and 1, respectively. Note that the difference between the three models lies exclusively in the factor structure; the conditional survival curves always factorize in their factor dependence.

4.2.3. *General Gaussian chain*

Here we keep the factor structure of the previous section, i.e.,

$$Z = (Z^{(1)}, \tilde{Z}^{(2)})\tilde{\Phi}_2^{\rho z}, \tag{4.5}$$

but we notice that the product form of (4.4) can be generalized by introducing a correlation parameter ρ_ϵ. We replace the product of the two normal distributions by a single bivariate to get

$$g_i(t, z) := \begin{cases} 1 - \Phi\left(\frac{h_i(t) - \sqrt{\rho_1} z^{(1)}}{\sqrt{1-\rho_1}}\right), & t \leq T_1 \\[2ex] 1 - \Phi\left(\frac{h_i(t) - \sqrt{\rho_2} \tilde{z}^{(2)}}{\sqrt{1-\rho_2}}\right) - \Phi\left(\frac{h_i(T_1) - \sqrt{\rho_1} z^{(1)}}{\sqrt{1-\rho_1}}\right) \\[2ex] + \Phi_2\left(\frac{h_i(t) - \sqrt{\rho_2} \tilde{z}^{(2)}}{\sqrt{1-\rho_2}}, \frac{h_i(T_1) - \sqrt{\rho_1} z^{(1)}}{\sqrt{1-\rho_1}}, \rho_\epsilon\right), & T_1 < t \leq T_2 \end{cases} \tag{4.6}$$

It is easy to see that (4.6) satisfies all requirements of conditional survival curves, but we note that for continuity at $t = T_1$ we need to set $h_i^{(2)}(T_1) = -\infty$ and can no longer maintain (4.1). This reflects the fact that (4.6) does not come from a standard specification from trigger variables and that in some sense the multi-periodicity has been "baked into" the forward model — in contradistinction to the case where the model is defined as the forward version of a one-period model.

Indeed, one can recover this model from trigger variables as in [1], but there has to be a set of trigger variables for each period. The singularity in the default threshold is a consequence of the "restarting" of the trigger variables for each period.[4] If let the number of periods increase for a fixed final maturity we get a model where the trigger variables restart every instant. Intuitively this will converge to the

---

[3]At least this applies for quadrature-based implementations — Monte Carlo is still possible.
[4]As is well-known, in order to have finite default intensity from the start, the default threshold has to go to $-\infty$ there.

(Gaussian) first-passage-time copula model, but in view of our observation above about singularities in the default barrier, this limit may be subtle.

Looking at (4.6) in more detail we see that for $\rho_\epsilon = 0$, the bivariate term factorizes and we get back the form in (4.4), except for the denominator $1 - \Phi(h_i(T_1) - \sqrt{\rho_2} \bar{z}^{(2)})$. In a sense this denominator is still there, but since $h_i^{(2)}(T_1) = -\infty$ it evaluates to 1. In a sense we can obtain all the Gaussian chains presented above as limits of this model[5] — but some of these limits may be singular.

5. Concluding Remarks

In this paper we have introduced a novel way of combining factor copula models into new factor copula models. This technique can be applied straightforwardly to the problem of fitting market quotes for tranches with different maturities with a single, consistent model. Such a model contains the loss distributions for all maturities and can serve as the initial condition for the dynamics of portfolio losses as well as allowing consistent pricing and risk management of tranches with arbitrary detachment levels and maturities.

Another application is to construct term structures of forward base correlations for standard tranche detachment levels. This allows consistent pricing of tranches with these detachment levels, regardless of maturity. However, the pricing of tranches with non-standard detachment levels faces the intractable problem of interpolating base correlations.

The chaining mechanism developed in this paper is very general and allows the construction of chains where the dependency of losses in different forward periods can be controlled. Such chain models have "more dynamics" than one-period copulas and may be useful for some pay-offs, e.g., forward starting tranches. However, this dynamics is quite restricted and the computational complexity quickly becomes unmanageable.

In a sense the developments in this paper shows the severe limitations of trying to extend the factor copula framework to include dynamics — for this a more direct approach, perhaps some concrete version of the "top-down" frameworks set out in [9, 10], is more promising. However, chain models may still have an important role to play in the modeling of portfolio loss dynamics: that of providing the initial conditions.[6]

Appendix A. Factor Copula Models

Consider a general (one-)factor model specified by

$$X_i = b_i Y + \epsilon_i, \quad i = 1, \ldots, N, \tag{A.1}$$

---

[5] As well as some not considered explicitly such as eg, the limit $\rho_Z = 0$, $\rho_\epsilon = 1$
[6] In this respect we note that by using a copula model to provide the initial conditions we can compute sensitivities to single-name initial data such as CDS spreads and defaults. This would not be possible if worked just with the initial loss distributions.

where the b_i's are positive; the ϵ_i's are independent real-valued random variables with distributions F_i; and the common factor Y is independent of the ϵ's and has distribution F_Y over the domain D_Y.

We can use this as a model of the default times τ_i by considering each X_i as a default trigger and making the identification of events

$$\{\tau_i > t\} \equiv \{X_i > h_i(t)\}, \tag{A.2}$$

for suitable non-decreasing (default threshold) functions $h_i : \mathbb{R}_+ \rightarrow \mathbb{R}$. The obvious conditional independence of the trigger variables then translates into conditional independence of the default times:

$$P(\tau_i > t, \tau_j > s | Y = y) = P(\tau_i > t | Y = y) P(\tau_j > s | Y = y), \quad i \neq j.$$

The conditional survival curves are given by

$$g_i(t, y) = P(\tau_i > t | Y = y) = P(X > h_i(t) | Y = y)$$
$$= P(\epsilon_i > h_i(t) - b_i y) = 1 - F_i(h_i(t) - b_i y) \tag{A.3}$$

Evidently the factor distribution F_Y and the conditional survival curves g_i contain sufficient information to reconstruct the joint distribution of default times and so could be taken as a definition of the copula model.

The specification (A.1) is general in principle, but in practice it is highly constrained since the distribution of Y will typically taken to belong to some parameterized family of distributions and there is no gradual transition from one family to another. If we use instead the RFL formulation introduced in [2][7]

$$X_i = a_i(Z)Z + \epsilon_i, \quad i = 1, \ldots, N, \tag{A.4}$$

where now Z is standardized Gaussian, then it is easy to see that if we choose a_i such that the distributions for $b_i Y$ and $a_i(Z)Z$ are identical, i.e.,[8]

$$a_i(z) = b_i F_Y^{-1}(\Phi(z))/z, \quad z \neq 0, \tag{A.5}$$

the two specifications give the same copula model. Since the factor loading functions of the RFL model are unconstrained[9] we see that the RFL specification provides a non-parametric "cover" class of all other factor models. Of course, the RFL model also allows us to specify asset-specific factor loading functions and is in this sense much richer.

Acknowledgment

I am grateful to Leif Andersen for discussions and suggestions.

---

[7]Note that we are here using a different, and somewhat more convenient, normalization.

[8]We need to make the technical assumptions that F_Y is invertible, i.e., has positive density, and has median $\frac{1}{2}$.

[9]Any positive factor loading function defines a valid model. However, many functions are equivalent in the sense that they define the same model. It may be useful to pick a unique representative of each equivalence class, for example by imposing the natural requirement that $a(z)z$ be increasing.

References

[1] L. Andersen, Portfolio losses in factor models: Term structures and intertemporal loss dependence, Working paper, Bank of America (2006).

[2] L. Andersen and J. Sidenius, Extensions of the Gaussian copula, *Journal of Credit Risk* **1**(1) (2004/2005) 29–70.

[3] L. Andersen and J. Sidenius, Cdo pricing with factor models: Survey and comments, *Journal of Credit Risk* **1**(3) (2005) 71–88.

[4] L. Andersen, J. Sidenius and S. Basu, All your hedges in one basket, *Risk* **November** (2003) 67–72.

[5] D. Guegan and J. Houdain, Collateralized debt obligations pricing and factor models: A new methodology using normal inverse gaussian distributions, Working paper, Ecole Normale Superieure (2005).

[6] J. Hull and A. White, The perfect copula, Working paper, University of Toronto (2005).

[7] A. Kalemanova, B. Schmid and R. Werner, The normal inverse gaussian distribution for synthetic CDO, Working Paper (2005).

[8] L. McGinty, E. Beinstein, R. Ahluwalia and M. Watts, Credit correlation: A guide, Technical paper, JP Morgan (2004).

[9] P. J. Schonbucher, Portfolio losses and the term structure of loss transition rates: A new methodology for the pricing of portfolio credit derivatives, ETH Zurich working paper (2005).

[10] J. Sidenius, V. Piterbarg and L. Andersen, A new framework for dynamic credit portfolio loss modelling, Working paper, RBS, BarCap and BofA (2005).

[11] S. Willeman, Fitting the cdo correlation skew: A tractable structural jump model, Working paper, Aarhus Business School (2005).